Seven Modern Plagues

and How We Are Causing Them

Seven Modern Plagues

and How We Are Causing Them

Mark Jerome Walters

ISLANDPRESS

Washington | Covelo | London

Library of Congress Cataloging-in-Publication Data

Walters, Mark Jerome, author.
 Seven modern plagues and how we are causing them / Mark Jerome
Walters.
      p. ; cm.
 Includes bibliographical references and index.
  ISBN 978-1-61091-465-9 (paperback : alk. paper) -- ISBN 1-61091-
465-1 (paperback : alk. paper)
 I. Title.
 [DNLM: 1. Disease Outbreaks--prevention & control--Popular
Works. 2. Epidemics--economics--Popular Works. 3. Environmental
Health--Popular Works. WA 105]
 RA649
 614.4'9--dc23

                    2013031297

⊕ Printed on recycled, acid-free paper

Manufactured in the United States of America
10 9 8 7 6 5 4 3 2 1

Keywords: mad cow disease, HIV/AIDS, salmonella, Lyme disease,
hantavirus, West Nile virus, SARS, bird flu, swine flu, MERS

To my dearest Noelle, Will, and Anna

Contents

Preface

The first edition of this book was published more than a decade ago, with warnings that the age of "ecodemics" had arrived. The years since have only confirmed the suspicion that human activity is behind many of them. Environmental change—whether brought about by agriculture, urbanization, or various technologies—has given rise to entirely new diseases and caused existing ones to expand their range.

New ecodemics alone would more than justify this updated edition. The first influenza pandemic since the Hong Kong flu of 1968 emerged in 2009; a new tick-borne infection appeared in the Midwest; a second novel bird flu was identified in China; and yet a new SARS-like virus recently appeared in the Middle East.

But in addition to emerging diseases, the past decade has produced a host of new information about the six plagues covered in the first edition:

. . . .

- Chronic wasting disease of deer, covered almost as a footnote in the first edition's chapter on mad cow disease, has greatly expanded its range—and apparent risks—in the United States.
- Genetic analysis has now pinpointed the origins of the human immunodeficiency virus (HIV).
- New fronts have opened up in the battle against increasing antibiotic resistance. (On the positive side, the US Food and Drug Administration recently began taking more decisive action on the issue.)
- Additional studies have confirmed the link between forest fragmentation and urbanization and the increased risk of Lyme disease.
- A major outbreak of hantavirus in California's Yosemite National Park in 2012 has renewed scientific interest in that infection.
- In 2012, more than 5,500 infections with West Nile virus were reported—mostly in Texas—making it the worst outbreak since 2003. Research during the past decade has greatly expanded the understanding of the ecology of this virus, including its connection to the American robin.

Disease is a process, not an event. Over the past ten years, the plagues have continued to progress, as has our knowledge of them. The first edition of this book now stands as a snapshot of where we were. This new edition offers a more complete understanding of where we are today and what the future may hold.

Introduction

I first learned of the strange new disease in the city nearly fifteen years ago while reading the *New York Times*. Just across the East River from my Manhattan office, several elderly victims had been admitted to Flushing Hospital Medical Center in Queens. They had been having trouble walking, were confused, and in some cases were comatose. Several soon died. Nearly a month passed before the affliction was identified as brain inflammation caused by an exotic virus. Before long we learned it was West Nile encephalitis, a disease originally seen in Uganda that was now being found for the first time in the Western Hemisphere.

Cases of the illness soon emerged near where I lived, in northern New Jersey, an hour's train commute from Manhattan. The idea that a potentially fatal disease almost unheard of there a few months before had suddenly popped up near my home was terrifying. Was this how the Black

Death, which wiped out as much as one-third of Europe's population in the 1300s, or the 1918–1919 Spanish influenza epidemic, which killed at least 20 million people in my parents' lifetime, began? As a veterinarian, I am familiar with diseases, including some frightening ones. But no amount of medical training had prepared me for new, life-threatening diseases heretofore unknown in my neighborhood.

At the time, I wanted to dismiss West Nile virus as anomalous. Problem was, it wasn't the first new disease to appear during my lifetime or even in my town—nor would it be the last. Some outbreaks seemed like faraway curiosities, whereas others had become personal, everyday concerns. Lyme disease, which hadn't even been described until the mid-1970s, was now endemic in Morris County, where I lived. And then there was HIV/AIDS, a disease whose deadly global spread was known to almost everyone, not least of all those of us in the New York City region. Even mad cow disease and other afflictions I knew of only through the scientific literature sometimes seemed only a supermarket or an ill airplane passenger away.

In late 2002, the point was brought home when word came from China that a previously unknown coronavirus had been causing a form of severe acute respiratory syndrome—later known as SARS. This deadly and highly contagious pneumonia was rapidly spread by international air travelers. Within a month almost twenty countries, including the United States and Canada, reported cases.

According to the Centers for Disease Control and Preven-

tion, the outbreak began in Guangdong Province in southern China when dozens of people there began to experience
headaches, muscle soreness, and dry coughs that quickly deteriorated into life-threatening pneumonia. Within months,
the illness had spread throughout Guangdong, where government authorities, fearing social unrest and the loss of
tourism, tried to keep the outbreak secret. No medical statistics were released, and journalists were prohibited from
reporting on the deadly epidemic.

In February 2003, Liu Jianlun, a sixty-four-year-old kidney specialist from Zhongshan Hospital in Guangdong,
traveled to Hong Kong, where he stayed in room 911 at the
Metropole Hotel. He had a fever and had not felt well for
five days, but when he began to have trouble breathing, he
went to Kwong Wah Hospital in Hong Kong. Suspecting he
had contracted the highly infectious illness, he asked to be
put in an isolation unit. He died several days later. China's
secret was no more.

By March 2003, Chinese authorities had begun to reveal
the extent of the outbreak in Guangdong. But then the disease was already sweeping through Hong Kong, infecting
hundreds and killing dozens. It had begun to invade Beijing
and other cities. It had also arrived in the United States.
The disease spread readily from person to person through
coughing, sneezing, and other means. It was also quickly
disseminated around the globe by jet aircraft. One traveler
flew from Hong Kong to Frankfurt and Munich, then on to
London, back to Munich and Frankfurt, and then again to

Hong Kong, apparently before even suspecting he had contracted a new disease.

In mid-March, the World Health Organization (WHO) declared the new virus "a worldwide health threat." Strict isolation of suspected cases and extreme precautions by health care workers eventually began to slow the spread in many countries, but the virus remained out of control for weeks in China, where dozens of new cases were reported every day. At the time, epidemiologists suspected that about 7 percent of patients died. Researchers soon learned that the fatality rate, at least in Hong Kong, was closer to 15 percent for those under sixty years of age and more than three times that for those older than sixty. Finally, in May 2004, WHO announced that through quarantine of infected patients and other control measures, the SARS outbreak had been contained.

Genetic analysis suggested that the SARS virus had come from a nonhuman animal. Scientists suspected that the coronavirus was spread by the masked palm civet cat, a weasel-faced tree-dweller native to Asia and consumed as a medicinal in China in the belief that it helps people withstand cold weather. (But bats may have been the original source.) Officials ordered the killing of every civet cat in captivity in southeastern China's Guangdong Province. The killing of the estimated 10,000 cats would be followed by efforts to trap and eliminate the animals in the wild.

For someone trained as a veterinarian, it is no surprise that diseases frequently jump from other species to humans.

Nearly 75 percent of new human diseases discovered over the past few decades are carried by wild or domestic animals. We acquired many ancient diseases from other animals, including smallpox from cattle and, apparently, the common cold from horses.

An enormous reservoir of potentially disease-causing viruses resides in wild animals, with many of these microbes remaining undetected until they suddenly appear on the human horizon. What's more, when a particular virus exists in both humans and other animals, as opposed to being present only in humans, there is almost no way to eradicate it. The best we can do is identify the animal reservoir and try to protect ourselves by showing a healthy respect for the natural boundaries between that species and us.

All this was not even to mention the super-exotic diseases such as Ebola hemorrhagic fever, which had undergone periodic irruptions among people and some wildlife in Sudan and Zaire during the previous two decades—deadly outbreaks that continue to this day. Infection with the usually fatal Ebola virus causes massive internal hemorrhaging. Barely a decade before West Nile virus broke out in New York, several monkeys infected with Ebola virus were imported into Virginia in what could have led to the first human outbreak of the disease in the United States. Fortunately, quarantine of the animals and rapid identification of the virus prevented its spread to the monkeys' human caretakers. Still, the evidence was clear: numerous new, sometimes fatal, infectious illnesses were pounding at the door. Some had already made it through.

. . . .

But hadn't the surgeon general of the United States proclaimed, way back in the late 1960s, that the time had come when Americans could "close the book on infectious diseases"? Hadn't the miracle of modern medicine all but ended the war against pestilence?

In fact, now, nearly four decades later, infectious disease still kills more than one in three people worldwide. The World Health Organization reported in 1999 that "diseases that seemed to be subdued . . . are fighting back with renewed ferocity. Some . . . are striking in regions once thought safe from them. Other infections are now so resistant to drugs that they are virtually untreatable." Even the Central Intelligence Agency has expressed concern about the resurgence of infectious disease. In 2000 the CIA predicted that emerging infections will "complicate U.S. and global security over the next 20 years . . . endanger U.S. citizens at home and abroad, threaten U.S. armed forces deployed overseas, and exacerbate social and political instability in key countries and regions."

This prediction was partially realized when, in April 2003, an estimated 10,000 residents of Chagugang, a two-hour drive from Beijing, rioted and gutted a building where SARS patients were supposedly to be housed. SARS riots elsewhere in China soon followed.

Scientists tell us that this global rise in infections comprises two general trends. Old diseases once believed to be controlled have resurged and in some cases have sprung up in new regions of the world. In recent years, malaria, an ancient disease, has dramatically increased in many areas, such as East Africa. This mosquito-borne illness kills nearly 2 million people annually. Half of the victims are children under five years of age. Some forms of the disease have become resistant to chloroquine, a mainstay of malaria treatment. The disease is also appearing in places where it was supposedly eliminated. In 2002, a fifteen-year-old boy and a nineteen-year-old woman in Loudoun County, Virginia, contracted malaria from mosquitoes near their home—the first time in at least twenty years that malaria had been found in both humans and mosquitoes in an American community. In some areas of the globe, the increase in malaria has been linked to a warming global climate and degradation of forests, which have given mosquitoes more places to breed.

In 2002 the tropical paradise of Maui, Hawai'i, reported its first case of dengue fever in more than fifty years. Transmitted by a mosquito bite, this virus causes a sudden high fever, severe headaches, joint and muscle pain, vomiting, and rash. It is sometimes fatal.

Perhaps like many people, I was tempted to dismiss these increases as artifacts of better detection methods. Weren't investigators simply picking up on diseases that had eluded our older, cruder methods of surveillance? Unfortunately, the facts do not support this optimism.

A second, equally ominous trend is the emergence of new diseases, of which WHO had identified more than thirty between 1980 and 1997—and many more by 2013. That list doesn't even include many earlier ones, such as rotavirus infection, the already mentioned Lyme disease (now the most common disease in the United States transmitted by a tick or other "vector"), Legionnaire's disease, Ebola virus and hantavirus infections, and toxic shock syndrome—to mention a few. In 1995 this plethora led the Centers for Disease Control and Prevention to create an entirely new journal, *Emerging Infectious Diseases*, which has since published 10,000 articles.

In 2009 swine flu swept the globe, the first pandemic since the Hong Kong flu in 1968–1969. Continuing to lurk in the background is the highly deadly (but not contagious, yet) bird flu, H5N1. In 2013 yet another variety of bird flu, caused by the similarly lethal H5N7 virus, emerged in China.

In 2011, two men on farms in northwestern Missouri came down with a severe illness that caused them to be hospitalized for a week with high fever, diarrhea, nausea, muscle pain, erratic blood counts, and liver problems. After intensive investigation by CDC and other scientists, a new disease carried by a tick was identified. The scientists who identified it called it Heartland virus.

The following year, in 2012, the SARS-like MERS-CoV virus burst onto the scene in Jordan and Saudi Arabia. And in 2013, the CDC released a report detailing the rise of new antibiotic-resistant bacteria in medical settings, which Dr. Thomas R. Frieden, director of the CDC, called "nightmare bacteria."

While antibiotics, better sanitation, and other measures

have lowered the percentage of deaths from infection world-wide since 1900, such improvements have hardly closed the book on infectious disease. If anything, we are in the process of writing entirely new volumes.

This emergent-disease phenomenon is actually more widespread than is at first apparent. Populations of frogs and other amphibians around the globe have declined dramatically since the 1980s, partly because of novel infectious diseases. Plagues are striking a wide range of other species, including crayfish, seals, honeybees, wolves, gorillas, prairie dogs, ferrets, penguins, snails, snakes, wild dogs, salamanders, pelicans, and kangaroos, to name a few. Infections threaten to drive some species to extinction. Ebola hemorrhagic fever is rapidly wiping out many of the world's remaining wild gorillas. A cancer epidemic, apparently caused by a virus, threatens many species of sea turtle worldwide. Chronic wasting disease, a brain-destroying affliction similar to mad cow disease, is spreading among wild deer and elk in the western United States and could eventually spread throughout white-tailed deer in the East.

We've all heard some of these accounts, but our understanding tends to be based on piecemeal news, with little sense of an encompassing story. In some ways we are getting the least important part of the picture. Media reports usually describe isolated battles against new diseases and rarely tell us the larger ecological story of which many new afflictions are a part. The larger story is not simply that humans and other animals are falling victim to new diseases; it is that we are causing or exacerbating many of these ecodemics.

Intensive modern agriculture, clear-cutting of forests, global climate change, decimation of many predators that once kept disease-carrying smaller animals in check, and other environmental changes have all contributed to the increase. This is not even to mention increased global travel and commerce, which can rapidly spread many diseases. This view is not an alarmist's leap of the imagination; it is quickly gaining ground as evolutionary and epidemiological fact. Noted scientist Peter Daszak, executive director of the Consortium for Conservation Medicine in Palisades, New York, put it this way: "Show me almost any new infectious disease, and I'll show you an environmental change brought about by humans that either caused or exacerbated it."

Environmental change and human behavior have long played a role in fostering epidemics. In fact, historians such as William H. McNeill believe that major, extended waves of epidemics have swept across the human species on several occasions, beginning some 10,000 years ago, when the first agricultural settlements and close human contact with cattle and other livestock gave microbes a new bridge for jumping to humans, aiding the rise of smallpox, measles, leprosy, and other diseases.

Then, some 2,500 years ago, increasing contact among established centers of civilization opened new avenues for emergence or spread of disease, giving rise to a second extended wave of epidemics. Increased global exploration then ushered in a third phase of epidemics as indigenous peoples in Africa, the Americas, the Pacific region, and elsewhere fell victim to introduced diseases.

Mercifully, throughout the late nineteenth and most of the twentieth centuries, many societies enjoyed a dramatic decline in infectious disease. This was largely because a state of relative equilibrium had been reached: societies had developed immunity to many of these old diseases and had adjusted their ways of life to control them. Unfortunately, this period of relative microbiological peace has been short-lived; humans now appear to be entering a fourth phase of epidemics, spawned by an unprecedented scale of ecological and social change.

One example of a human-assisted, if not human-made, disease is Nipah virus, named after the place of its discovery in Malaysia in 1999. For humans, the infection is often mild, though it may elicit influenza-like symptoms, including high fever and muscle pain. In some instances, the disease progresses to encephalitis, or inflammation of the brain, leading to convulsions, coma, and, in half of symptomatic patients, death.

The natural reservoir of Nipah virus is the giant fruit bat of Southeast Asia, to which the virus apparently causes no harm. These enormous "flying foxes" normally feed in wild fruit trees in sparsely settled areas. But since the 1980s, logging and the spread of agriculture in Southeast Asia have decimated the bats' forested homes. In 1997 and 1998, human-set forest fires in Borneo and Sumatra, spurred by an

El Niño–linked drought, blanketed much of Southeast Asia with a thick haze. This confluence of fires, unusual drought, and forest degradation caused the natural fruit crop to fail and forced the flying foxes to migrate farther north in search of food. They ended up in new locales, such as cultivated fruit orchards near pig farms in Peninsular Malaysia, where their hitchhiking virus attacked new species whose immune systems were unprepared for this microscopic invader. More than a hundred people died, and the Malaysian pig industry was devastated. Although reappearances of the virus in the region have been sporadic, the high fatality rate makes it a serious public health threat.

People in the United States, Europe, Japan, and other developed countries can no longer safely relegate new, exotic, or deadly diseases to faraway places. HIV/AIDS has invaded every part of the globe. The human form of mad cow disease infected people through meat served on ordinary dinner tables throughout England. In some regions, Lyme disease and West Nile encephalitis pose a risk to people in the city, in suburbia, or on a casual walk in the woods in spring or summer.

In the pages that follow, I tell of the human role in fostering modern epidemics through the stories of seven modern diseases. (The number of diseases in this book alone has grown, up one from the six covered in the first edition.) Mad cow disease, first isolated in cattle in England in 1986, causes an ultimately fatal degeneration of the victim's brain. HIV/AIDS is moving toward the top of the list of humankind's most deadly modern scourges—with no end in sight. A form of food poisoning caused by a new strain of the salmonella bacterium known as DT104, which people usually contract by eating contaminated meat, is resistant to almost all the antibiotics commonly used to treat more traditional forms of salmonella food poisoning. The rate of Lyme disease is increased by ecological disruption of forests. A fatal illness caused by a newly discovered hantavirus, a virus spread by mice, makes regular appearances in the American Southwest and other areas, depending, in part, on global meteorological cycles. Unknown to almost all Americans just a few years ago, West Nile encephalitis is becoming a dismal fact of life as the virus spreads throughout most of the United States. In addition to the original six, this edition explores the story of a seventh plague for the ages—pandemic influenza—including both swine and bird flus. Seven diseases, seven parables of the unintended consequences of disturbing the natural systems on which our own health depends.

The ecological whole of these seven diseases is far greater than the sum of their individual parts, and their significance is far greater than the relatively low incidence of some of these diseases. Together, these epidemics offer insights into the way we live, how we think, and the assumptions we embrace as children of the age of medical miracles. For all that modern medicine has prolonged life and relieved suffering, it has also fed the profoundly dangerous illusion that we are above or apart from the natural world, with its weather, forests, and cycles of life and death. The seven diseases described in this book remind us that no amount of medical technology can rescue us from the heart-stilling fact that human beings, as William H. McNeill put it, "will never escape the ecosystem and the limits of the ecosystem. Whether we like it or not, we are caught in the food chain, eating and being eaten. It is one of the conditions of life."

This realization is not all bad. In preserving the ecosystems on which health fundamentally rests, we stand to protect the health of many people for generations to come. In carelessly exploiting water, forests, fossil fuels, other species, and other natural resources, we will continue to sacrifice the long-term physical health of many for the financial gain of a few. Preservation of natural ecosystems, along with greater social equity, research, good surveillance, and benefits of modern medicine, can improve the health not only of people but also of many other species. Human health does not belong to us alone. Nor, unfortunately, do the plagues we are all now experiencing.

The Dark Side of Progress:
Mad Cow Disease

1.

Near the village of Midhurst, West Sussex, an hour's journey south from London through green glens and soft hills,
stands a seventeenth-century brick-and-timber farmhouse
surrounded by purple hydrangeas and lipstick-red geraniums that tilt in the breeze. The lichen-covered clay tile
roof and weathered walls seem to have grown from the
earth itself. Sprays of red and yellow flowers spill from every corner of the grounds, and wild roses climb a trellis
above a gate leading down to lush pasture and an ancient
stone stable. It is as if Pitsham Farm were drawn from
the enchanted poetry of William Wordsworth, where "majestic herds of cattle, free / To ruminate, couched on the
grassy lea." Or so it might have seemed until, three days

before Christmas in 1984, one of Peter Stent's cows began acting strangely.

"At first we dismissed her as a cow with a bad disposition, kicking in the milking parlor and all that," Stent told me. But when she got worse, Stent called his veterinarian, David Bee, who visited the farm. The cow hunched her back, leading Bee to believe she might have a painful kidney ailment. More cows soon fell ill, and Bee returned several times to attempt to diagnose the ailments. The first cow grew worse, developing head tremors and an unsteady gait. In February 1985 she died. The mysterious illness continued to spread through the herd. At a loss for a diagnosis, Bee dubbed the affliction "Pitsham Farm syndrome." Whatever the root of this malady, Bee concluded that it was attacking the brain, and he and Stent decided to ship a sick cow to the local agriculture ministry.

"I shall never forget that cow," Stent said. "The man came with a trailer already loaded with two sheep on their way to slaughter. When we prodded the cow into the trailer, she saw the sheep; then she went berserk and killed them. I thought she was going to destroy the whole trailer. She was extremely violent." Unfortunately, when the cow arrived at the local ministry she was killed with a gunshot to the head, which destroyed the brain and rendered it useless for analysis.

Determined to find the cause, Stent and Bee loaded up cow number 142—the tenth cow to be afflicted with the illness—and had her driven to the ministry. The head was removed intact and sent to the Central Veterinary Laboratory in Weybridge, Surrey, where the brain could be examined by a pathologist.

A stocky man with gentle blue-gray eyes, Stent sat in a lawn chair at a table and paused to sip his tea. A row of royal purple foxgloves nodded in an early summer breeze from the English Channel, twenty miles away. "Spooky behavior for these kindly animals," he recalled. "I cared about them and hated to see them sent to slaughter."

Stent's wife, Diana, appeared in the sunny yard and re-filled our cups. A wood thrush sang three platinum notes followed by a reedy tremolo from a bush near an abandoned brick privy. Stent separated his right hand from his teacup long enough to make a short, sweeping gesture. "It's becoming more difficult to make a living from the farm anymore. I'm fortunate indeed to have other means. The price of milk has gone so low that we can't compete with larger operations. Now, with the Channel Tunnel open, tanker trucks bigger than my milking parlor bring cheap milk from the continent. We have 600 beef cattle, but people's feelings have really changed about eating meat."

Eager to give his respected veterinarian a place in our conversation, Stent called Bee on his cell phone to arrange a meeting. We drove the backroads of the 600-acre farm past mostly empty pastures. When we arrived at the dilapidated milking parlor, Stent leaned out the car window and pointed inside the building's wide doorway. "I couldn't justify mod-ernizing the operation in light of things. Now, look in there. Those are the old feed bins. At the time, I couldn't imagine what my cows were being fed. It's in there we first noticed the cows acting strangely. That's the spot where BSE began as far as the history books are concerned," he said, using the initials for bovine spongiform encephalopathy, the technical name for mad cow disease.

Bee's clinic was in the village of Liss, a twenty-minute drive. "As if BSE weren't enough, the foot-and-mouth epidemic last year finished off a lot of farms," Stent said as we drove along. He was referring to another highly contagious cattle disease that had recently swept through the United Kingdom. Although not dangerous to humans, it is one of the most contagious and economically devastating livestock diseases. "We didn't get foot-and-mouth at Pitsham, but we were quarantined like farms throughout the UK. Any farm with the disease, all the animals were burned."

Bee greeted us in the waiting room of his clinic and ushered us into a treatment room so we could talk without interruption. A man in his late forties or early fifties, he wore wire-rimmed glasses, and his eyes shone with inquisitiveness. "Still haunts me sometimes," he said, recalling his first encounters. "You'd never recognize it in an undisturbed grazing herd. Then I'd walk up to the fence and suddenly a cow two hundred yards away would lift its head and fix its gaze on me with an eerie hypervigilance. 'That cow's infected,' I'd say to myself. If you stressed it, its symptoms could explode into kicking, tremors, aggression, a wobbly gait. An infected cow would come apart at the seams. Really spooky."

The task of examining cow 142's brain fell to Carol Richardson, a pathologist at the Surrey laboratory. She noted a strange sponge-like appearance strikingly similar to what is found in sheep with a well-known neurological disease called scrapie. Richardson wrote "spongiform encephalopathy" on the necropsy form and left the slide for Gerald Wells, her supervisor, to examine. Wells confirmed Richardson's diagnosis and filed the slide.

A year later, a cow from Kent developed similar symptoms; it became clear that the disease was not limited to Stent's farm. When this cow's brain reached Wells's laboratory, he discovered that it also had a spongiform encephalopathy. In 1987, fourteen months after Richardson's diagnosis of cow 142, Wells hailed his own discovery of "a novel progressive spongiform encephalopathy in cattle" and published a paper on his finding without so much as mentioning Richardson's diagnosis of the cow from Pitsham Farm—the first-ever documented case of mad cow disease.

Prior to Wells's 1987 publication, several cows at a farm in Malmesbury, Wiltshire, seventy-five miles east of Pitsham Farm, had also developed a fear of walking over concrete or venturing around corners. Some hung their heads low as if exhausted. Others developed a high-stepping gait in their back legs as if walking on hot pavement. Milk production dropped. Cows fell down and couldn't get up. The epidemic soon affected fourteen counties in southern England.

Although mad cow disease was apparently a new affliction, it belonged to a class of known brain-wasting diseases called TSEs, or transmissible spongiform encephalopa-

thies. The name indicates that the diseases can be contagious and lend a spongy appearance to the brain, just
as in the cow's brain Richardson had described. The first
human TSE, Creutzfeldt-Jakob disease (CJD), was described in the 1920s. This degenerative disease leaves its
victims, in the early stages, with loss of memory, unsteady
gait, muscle spasms, and jerky, trembling hand movements.
Another TSE, scrapie in sheep and goats, was scientifically
described in 1936, although one of its symptoms—violent
scratching to the point of mutilation—had been known for
centuries. About a decade later, a TSE was identified in
ranch-reared minks.

In 1957 yet another human TSE, kuru, was identified in
Papua New Guinea. Then, in 1967, chronic wasting disease
was identified in some deer and elk in the western United
States. BSE was officially added to the list of TSEs in 1987
with the publication of Wells's paper. But mad cow disease
was not just another TSE: never before had the affliction expressed itself in such a widespread outbreak. By 1988 more
than 2,000 cows had been stricken, and in 1992 alone more
than 35,000 cases of BSE in cattle would be reported. By
January 1993 almost 1,000 new cases in cows were being
reported *every week*. "Incurable Disease Wiping Out Dairy
Cows" proclaimed a headline in London's *Sunday Telegraph*
in 1987: "A mystery brain disease is killing Britain's dairy
cows, and vets have no cure." Farmers began to fear for their
livelihoods and their rural traditions. But at least they were
not fearing for their own lives—not yet.

2.

In October 1989 a report surfaced describing a woman, believed to be at least thirty-six years old, who had been diagnosed with Creutzfeldt-Jakob disease. That disease struck, according to one study, fewer than one in every 10 million people in Britain and Wales each year, and its usual victims were middle-aged or older people; the average age of victims at the onset of the disease was fifty-seven. CJD in the young—a teenager, for example—is so rare as to typically occur only once every twenty or thirty years. Conventional wisdom held that CJD was either inherited or contracted from contaminated surgical instruments, transplants, or cadaver-derived growth hormones once used to treat dwarfism. When it was learned that the young woman had been associated with a farm where mad cow disease was present, people began to wonder whether she had contracted her disease from an infected cow. This was dismissed by a government scientific committee, however, which concluded that "the risk of Bovine Spongiform Encephalopathy to humans is remote."

In August 1992 came the case of Peter Warhurst, a sixty-one-year-old dairy farmer at Meadowdew Farm in Simister, north of Manchester, who died of Creutzfeldt-Jakob disease. Warhurst had culled a "mad cow" from his herd three years before. The prestigious British medical journal *Lancet* described this as "the first report of CJD in an individual with direct occupational contact with a case of BSE." The report said that the case was probably a chance occurrence but

. . . .

raised "the possibility of a causal link." It was not a link the government wanted to hear about. Livestock is a mainstay of the United Kingdom's economy, and the stakes were huge.

Kevin Taylor, the government's assistant chief veterinary officer responsible for BSE control, publicly dismissed the notion of a link between mad cow disease and CJD, saying, "I don't think that a link between this case and BSE is even conjectural." This echoed repeated claims by British agriculture minister John Gummer that there was "no evidence anywhere in the world of BSE passing from animals to humans" and that "on the basis of all scientific evidence available, eating beef is safe." At a boat show in Ipswich in 1989, Gummer had vouched for the safety of beef, this time in a BBC television report that showed him helping his four-year-old daughter, Cordelia, chomp down on a beef burger nearly the size of her face. "When you've got the clear support of the scientists who deal with these matters [and] the clear action of the government, there is no need for people to be worried," he proclaimed, "and I can say completely honestly that I shall go on eating beef and my children will go on eating beef because there is no need to be worried."

But new cases kept emerging. In May 1993, Duncan Templeman, a sixty-four-year-old Somerset dairy farmer, came down with CJD. There had been three cases of BSE on his farm, and he was a beef-eater. Eight months later, in January 1994, a third dairy farmer, from Just, in Cornwall— some of whose cows had also contracted BSE—entered a hospital with loss of memory and slurred speech. The fifty-four-year-old farmer soon became mute, and he died of pneumonia some months later. The *Lancet*, which reported the case, concluded that "the occurrence of CJD in another dairy farmer . . . is clearly a matter of concern." Although the report emphasized that the farmer might have contracted the disease from his cows, the government's Spongiform Encephalopathy Advisory Committee emphasized that he might *not* have done so and the case therefore did not require "the Government to revise the measures already taken to safeguard public health against occupational and other possible routes of exposure to the BSE agent." But one member of the committee warned that should a fourth case arise, the tide of probability would turn: farmers were probably catching CJD from their cows. In September 1995, a fourth ill farmer came to light. As if that weren't convincing enough, a rash of puzzling CJD cases had begun to occur in beef-eating young people not associated with farms.

In 1993 fifteen-year-old Victoria Rimmer of Connah's Quay, Deeside, came down with CJD—the youngest reported victim in Great Britain in almost twenty-five years. Victoria had been exceptionally healthy until May 1993, when she began losing weight, developed trouble with her vision, and soon became apathetic. A brain biopsy revealed spongiform encephalopathy. Her condition deteriorated. She had fits, her body twitched uncontrollably, and she went blind. According to her mother, the British newspaper *Today* reported, beef burger was Vickie's favorite food. Kenneth Calman, England's chief medical officer, countered,

"No one knows what illness she is suffering from . . . there is no evidence whatever that BSE causes CJD." Victoria soon fell into a coma that lasted four and a half years, ending in her death.

The notion that CJD might be linked to a person's diet was not new, and the supporting evidence was as tantalizing as it was scant. In 1984 the *American Journal of Medicine* reported four cases in which individuals who commonly ate animal brains—those of wild goats, squirrels, and pigs— came down with CJD. The authors concluded: "Our case, along with experimental evidence for oral transmission of Creutzfeldt-Jakob disease and other spongiform agents, sup- port[s] the hypothesis that ingestion of the infective agent may be one natural mode of acquisition of Creutzfeldt- Jakob disease."

After the identification of mad cow disease, not surpris- ingly, such speculation increased. In 1997 a neurologist at the University of Kentucky came across a CJD patient in Florida, a native of Kentucky who had a long history of eat- ing squirrel brains back home—not an uncommon practice in rural parts of the state, where the brains are sometimes scrambled with eggs or put in a meat and vegetable stew called burgoo. The neurologist later discovered that all five patients of a neurology clinic in western Kentucky who were suspected of having CJD had a history of eating squir- rel brains. The patients were not related, and they all lived in different towns, facts that minimized heredity or direct contact as a means of transmission. The study was widely

reported in the media but criticized in the scientific community; for one thing, squirrels apparently don't get spongiform
encephalopathy.

In 1998 the *Lancet* reported the intriguing case of a sixty-
year-old man from Italy who had been admitted to the hospital with muscle contractions, an unsteady gait, visual diffi-
culties, and problems speaking. Two weeks after admission,
he became mute and couldn't swallow, and several months
later he died. The man, as far as anyone knew, had no unusual eating habits. But about the same time he was admitted to the hospital, his seven-year-old cat developed uncon-
trollable twitches and episodes of frenzy and hypersensitiv-
ity to touch. The cat grew progressively worse and soon was
unable to walk. There was no evidence that the cat, which
slept on the owner's bed, had ever bitten him. Analysis of
cells from the man's and the cat's brains showed remarkably
similar abnormalities. Either the man caught CJD from his
cat, the cat caught it from the man, both were infected by a
common source, coincidence led them to become infected
independently, or the cases were simply misdiagnosed.

Epidemiologists rightly caution that for every victim of
CJD who had eaten the brain of a wild animal, there were
thousands of other people who had eaten the same thing
without contracting the disease. Such is the slippery nature
of anecdotal evidence. But it is also worth noting that of
the thousands of people who may have eaten BSE-infected
beef, only a select few contracted the human form of mad
cow disease.

By the end of 1995, ten suspected cases had been documented in young people in the United Kingdom. Senior
government officials continued to insist there was no link
with beef. Even as official denials flew, several prominent
scientists, including some government advisors, were preparing a paper for the *Lancet* that would confirm people's
worst fears—that the bovine disease and the human disease
were connected—by acknowledging the "possibility that
[these cases were] causally linked to BSE." Not until just
before the study's publication in the April 6, 1996, issue did
the British secretary of state for health, Stephen Dorrell, ad-
mit to the House of Lords that the ten young people probably were suffering from what had become known as variant
CJD, the human form of mad cow disease. Researchers soon
added physical evidence to the statistical case: the agent of
mad cow disease in humans was indistinguishable from the
agent that caused BSE.

Mad cow disease seemed like medical science fiction.
One of humankind's most ubiquitous domesticated com-
panions, the dairy cow, widely known for its gentle nature,
and a frequent subject of poetry and painting, had delivered
a ferocious new disease unto its keepers. No one could say
how the cows had gotten it, but speculation soon shifted to
their pastured brethren the sheep. It was one more connection in a strange set of circumstances that seemed to link
sheep, cows, and humans in a bizarre and unprecedented
web of affliction.

3.

Scrapie, the illness of sheep and goats that can cause the animals to madly scratch themselves raw, was first clinically recognized in Great Britain in 1732. An early description from Germany describes how suffering animals "lie down, bite at their feet and legs, rub their back against posts, fail to thrive, stop feeding, and finally become lame. . . . Scrapie is incurable. . . . A shepherd must isolate such an animal from healthy stock immediately, because it is infectious and can cause serious harm in the flock." The French term for the disease translates as the "malady of madness and convulsions."

Not until 1936 was scrapie proven to be infectious, though its origins remained a mystery. In 1966 researchers at Hammersmith Hospital in London suggested it was no ordinary infectious agent because, whatever it was, it possessed no genetic material, or DNA. It therefore was not a living agent at all. Researchers drew their dramatic conclusions from the fact that DNA is fragile and can usually be destroyed by ultraviolet light, heat, or chemical disinfectants. But the scrapie agent remained infectious even after prolonged boiling, exposure to the extreme dry heat of sterilization, blasting with high levels of ultraviolet radiation, or even soaking in formalin and alcohol. Scrapie thus joined the strange fraternity of infectious brain-wasting diseases caused by a nonliving infectious agent. These are the perfect agents of disease: you can't kill them because they're already dead. But scrapie would not be the last in its class.

In 1957 American scientist D. Carleton Gajdusek and an

Australian colleague began investigating kuru, the fatal neurological disease that was killing the Foré people, an ancient tribe of about 15,000 in Papua New Guinea. The victims' brains looked so much like those of scrapie-infected sheep that in 1959 American veterinarian William Hadlow suggested the two diseases were the same. Like scrapie, kuru was infectious—in this case, it was passed through the tribe by the ritualistic eating of brains of the dead. A neuropathologist noted further that the brains of kuru victims looked a lot like those of CJD victims. If kuru looked like scrapie and CJD looked like kuru, then CJD looked like scrapie. These three brain-wasting diseases came to largely define TSEs, and for his work on kuru Gajdusek would receive the 1976 Nobel Prize in Physiology or Medicine.

But "TSE" was merely a descriptive term signifying something transmissible that made the brains of its victims spongy. It revealed little about the disease-causing agents themselves. In the early 1980s Stanley B. Prusiner, at the University of California, San Francisco, School of Medicine, sought to unravel the mystery. He proposed the heretical idea, for which he would later receive the Nobel Prize, that TSEs were caused by a special protein—that is, nothing more than one of the body's common molecular building blocks bound together in a lethal way. Unlike bacteria and viruses, these special proteins, which he called prions, do not reproduce—or at least not in the case of mad cow disease. Rather, once in the victim's body, they force normal proteins into abnormal configurations. Prions don't replicate;

they enslave. The notion of an infectious protein was strange enough. That it could also be inherited, as some cases of CJD showed, revealed an entirely new and fearsome type of infectious disease.

Just as the brains of various TSE victims looked a lot alike, so did some of the symptoms. In fact, the human form of mad cow disease was almost indistinguishable from the better-known CJD. The most striking clinical difference between the two diseases was the age of the victims. The term "variant CJD," or vCJD, was soon coined to reflect the finer distinctions.

The knowledge that people could get the disease from eating parts of infected cows, though a landmark discovery, was only one element of the larger story. How, in the first place, had the cows gotten it?

Scrapie-infected sheep remained the top suspect. Yet for centuries cows in England had intermingled with infected sheep, and there was not a single documented case of a cow becoming sick from scrapie. Nor, in the more than 350 years that scrapie had been known in England, was a single case documented of a person becoming sick from the sheep disease. If mad cow disease did in fact come from sheep, why had it just recently begun showing up in cows, let alone in people?

Perhaps a random mutation of the scrapie agent had suddenly made it infectious for cows—and people. Or maybe mad cow disease had nothing to do with sheep. Perhaps a protein in a cow's brain had randomly mutated into a lethal

TSE protein. Then again, conceivably mad cow disease had been around for a long time at such low frequency that it had never been detected in bovines, let alone in humans— until something happened to cause an explosive epidemic.

No one could say exactly what changes had caused the emergence of mad cow disease, but scientists soon began to wonder whether the intensive management practices in the production and husbandry of cows and sheep in the United Kingdom were responsible. Over the previous few decades, for example, as livestock production had intensified, many relatively small farms had been absorbed into huge industrial enterprises where livestock was treated like oil, natural gas, or any other commodity. The animals' natural needs for space, proper diet, and other comforts had been overshadowed by demands for greater efficiency and profit—but at an unexpected cost.

The fact that BSE seemed to be transmitted by consumption of certain parts of infected animals would have, under natural circumstances, prevented its spread between sheep and cows for the simple reason that these placid herbivores don't eat each other.

Or do they?

4.

Cows, sheep, and other herbivores evolved over millions of years to eat plants. Just about everything about them is geared to living in a world of greenery. Their teeth are designed for grinding tough plants, not for grabbing prey or cracking bones. Their large, padded lips help them grasp and pluck short grasses from the ground. Their broad cloven hooves help steady their weight on grass and soft earth, where lush forage is likely to be found. What's more, the bovine digestive system is designed to extract hard-to-get-at nutrients from grasses and other vegetation. Through grinding action and fermentation, their three "stomachs" break down and absorb the nutrients contained in their tough, lignin-based diet. Bacteria living in their gut are equipped to break down plant fibers. Cows, like all species, tend to function best within dietary boundaries drawn by evolution.

Violating such evolutionary boundaries can seem unnatural, if not disgusting. The term "rendering" is a euphemism for refining and repackaging animals' blood and guts into palatable feed for livestock. For example, for decades renderers in France routinely added human excrement to the mix, creating a high-protein feed supplement that was sold to livestock producers throughout Europe—a practice not stopped until the year 2000. Ignoring natural dietary boundaries of species is more than bad manners; it can also be bad for our health.

In the mid-twentieth century, meat producers realized they could save money if they recycled and sold the normally discarded by-products of butchered livestock, including the intestines, bladder, udder, kidneys, spleen, stomach, heart, liver, lungs, and other organs, as well as the bones. Through the process of rendering, these leftovers could be turned back into feed for cattle, sheep, and other herbivores. Nature's plant-eaters could be transformed into human-made carnivores.

The problem was that high-protein diets can cause serious problems in digestive systems designed for grass and other low-protein food. But the livestock producers saw this as nature's problem, not their own. Although cattle fed high-protein diets did routinely suffer from digestive problems, the animals usually survived to market, and, whatever the consequences for the animals, the effect on the profit margin was positive.

In the process of rendering, the use of heat, mechanical pressure, and chemical solvents reduces entrails and other organs into two basic chemical components. One product is fat, known as tallow, which is used for anything from soap manufacture and human consumption to production of animal feed and chemicals. The other product is greaves, used in fertilizer or as high-protein feed for cows, sheep, and other animals. Greaves can be further processed to yield a solid residue and small amounts of a valuable, highly purified fat used in perfumes and cosmetics. The solid residue can be ground up to produce concentrated meat and bone meal, or MBM. This is added to animal feed to boost the protein content, which can help the animals gain weight faster.

But scientists were puzzled. If rendering had caused mad cow disease, why had it not occurred forty years earlier, when rendering became a standard practice?

Although most prions survived rendering, one theory suggests that lowering the amount of heat or solvents in order to offset rising costs during the energy crisis of the 1970s had allowed even greater quantities of the scrapie agent to remain intact. Therefore, more prions ended up in MBM and in the diet of cows. Or, possibly, the modified rendering process physically altered the agent, making it more infectious for cattle. But such changes in rendering during the 1970s had occurred throughout Europe, so why did mad cow disease emerge only in the United Kingdom? The problem with the theory is that in subsequent experiments the use of solvents had little or no impact on the prion.

A unique British contribution to the emergence of mad cow disease may have been the dramatic increase in the number of sheep in the United Kingdom about the same time the disease emerged—from about 31 million sheep in 1980 to more than 44 million in 1990. This in turn meant that a greater number of scrapie-infected sheep carcasses were being sent to rendering plants—and ending up as MBM. By 1985 there were about two sheep for every cow in England, which meant that cattle in England were probably eating more scrapie-infected sheep, via MBM, than anywhere else in Europe. Perhaps the increased number of infected sheep consumed by cows tipped the balance to an infective dose of scrapie. But meat meal and bone meal with sheep material had been flowing to the feed bins of cattle in Britain for as long as seventy years. Surely if this were the route, cows would have become infected before the 1980s.

Or perhaps, some scientists theorized, the greater number

of scrapie-infected sheep simply increased the probability of a random change occurring in the infective agent, thereby turning a sheep disease into a bovine and human one. But there is no evidence for this—any more than there is for mutations in cattle leading to the outbreak of BSE.

"The most widely accepted hypothesis is that BSE originated in scrapie-infected sheep, but it's still just a hypothesis," Marcus G. Doherr of the Department of Clinical Veterinary Medicine at Switzerland's University of Bern, told me several years ago. "I don't think that riddle will ever be completely solved." Wherever it started, BSE rapidly spread in the cow population via feed containing meat and bone meal of infected animals. Whether farmers knew it or not—and Peter Stent was one who didn't—virtually all of them in the United Kingdom were feeding animal protein to their animals to make them grow faster.

Whatever the actual origin of the mad cow prion, intensive agriculture dramatically multiplied the agent and quickly sent it throughout the industrial food web. The British government's inquiry into the epidemic concluded that "BSE developed into an epidemic as a consequence of an intensive farming practice—the recycling of animal protein in ruminant feed. This practice, unchallenged over decades, proved a recipe for disaster." A 1988 ban on the feeding of recycled animal protein in the United Kingdom slowly stemmed the epidemic a few years later, but not before harm had been done to many humans and herds of cattle.

In January 1993, by the time the epidemic reached its

peak, an estimated 1 million cows had been infected. By November 2000, cases had been confirmed in more than 35,000 herds in the United Kingdom. Cases also appeared in Belgium, Denmark, Switzerland, Italy, Greece, Germany, France, the Netherlands, Portugal, Ireland, and Spain. Before long Japan and Israel had cases as well.

By early 2002, a total of 125 cases of the human form of mad cow disease had been reported worldwide: 117 in the United Kingdom, 6 in France, and 1 each in Ireland and Italy. Almost all the victims had lived in the United Kingdom between 1980 and 1996, the time of the BSE outbreak. In fact, there has never been a human case in which the patient did not have a history of exposure in a country where the disease was occurring in cattle.

5.

State of Colorado
Roy Romer, Governor
Department of Natural Resources
Division of Wildlife

20 December 1997
Christopher Melani

Dear Mr. Melani,
Thank you for participating in the Division of Wildlife's
chronic wasting disease survey. I am writing to let you know
that preliminary laboratory results indicate that the buck

you harvested . . . was probably affected by chronic wasting disease. When it's convenient, I would enjoy speaking with you to get some additional information on the location where this buck was harvested and discuss any observations you may have made in the field. In addition, I would appreciate your marking the harvest site on the enclosed map and returning it to us for inclusion in our survey database.

Although there is no evidence linking chronic wasting disease to human health problems, I also want to advise you that the Colorado Department of Public Health and Environment has recommended against consuming remaining meat from this carcass since laboratory tests indicate the animal may have been infected. If you wish to receive a refund on your license fee, please complete the enclosed application and return it to the Division of Wildlife. . . .

Sincerely,
Michael W. Miller, DVM, Ph.D.
Wildlife Veterinarian

The problem was that Chris Melani and his family had already eaten the venison. They'd also had part of it made into sausage, which they'd sent as presents to their friends.

Chris had shot the animal some time before. As requested by the state, he sawed off the head and took it to the fish and game office, where pathologists would examine the brain for the telltale sponge-like appearance of TSE. The state encouraged the hunting of deer in affected areas, hoping to reduce the number of infected deer. If a brain was found to have TSE, the state would notify the hunter within ten weeks, with instructions to dispose of the venison.

"I didn't get a notice, so I figured everything was okay with the deer," Chris was quoted as saying, having waited several weeks without receiving word from the state. His wife, who also ate some of the venison, said, "What's done is done. You just go on with your life." So far, they are still healthy. Given the extremely long incubation time for the disease, many more years may pass before they feel truly safe.

First described in 1967 in captive mule deer in northern Colorado, chronic wasting disease, or CWD, causes infected deer and elk to develop a blank stare and to walk in repetitive patterns while their bodies waste away, though not until 1977 was the infective agent determined to be a prion. By the mid-1980s CWD had spread to new parts of Colorado and into Wyoming. And despite the optimistic predictions of wildlife managers, by 2001 the disease had spread to Nebraska, to the Canadian province of Saskatchewan, and to captive herds in Oklahoma and Montana.

Even though there is no proven link between CWD and human illness, Chris Melani and his wife learned that three young venison-eaters had in fact come down with a degenerative brain disease with symptoms almost indistinguishable from those of mad cow disease. Twenty-seven-year-old Jay Dee Whitlock II of Oklahoma, a truck driver, began having difficulty finding his hometown. Soon he forgot how to drive his truck. He died of vCJD on April 7, 2000. A hunter, he had eaten deer meat regularly for most of his life.

In May 1998, twenty-eight-year-old Doug McEwan of Syracuse, Utah, also an avid deer hunter, had trouble calculating his travel expenses after a routine business trip and then began forgetting the names of his wife and other close relatives and his home telephone number. He soon had difficulty talking, writing, naming objects, and dressing. He died ten months after the onset of his illness. His brain showed the telltale sponginess of vCJD.

About a year before McEwan fell ill, a twenty-eight-year-old woman showed up at the emergency room several times with weakness and difficulty in walking. When her condition grew worse, she was admitted to the hospital, where she died. A brain autopsy showed spongiform encephalopathy. As a child, she had regularly eaten deer meat harvested by her father.

Given that there are normally only five reported vCJD cases per billion people each year in people thirty years of age or younger in the United States, three in such a short time was more than worrisome. According to the Centers for Disease Control and Prevention, of the 4,700 reported "classic" CJD deaths in the United States between 1979 and 1998, the victims' median age was sixty-eight.

"You cannot say with absolute certainty that CWD won't transmit to people, but there is no evidence that it will," veterinarian Tom Thorne, director of the Wyoming Game and Fish Department, told me. Thorne, who had hunted and eaten venison from the infected area for thirty years without ill effect, called himself the "longest ongoing experiment. All evidence indicates it doesn't jump from deer to elk to other animals, let alone to humans. You're more likely to get run over by a Winnebago in downtown Omaha."

Thorne emphasized that "the only thing that BSE and chronic wasting disease have in common is that they're both TSEs, and, possibly, they both originated from scrapie in sheep." Those similarities may be enough to stop some people from eating venison, but it obviously hadn't stopped Thorne. "CWD just isn't a risk to humans," he said. At the same time, Thorne acknowledged that it was "all a big mystery. We don't know where CWD came from. Maybe it was a spontaneous change in a protein that started in deer. Maybe they picked up the scrapie agent and it evolved to be pathogenic in deer. We just don't know." Pierluigi Gambetti, director of the National Prion Disease Pathology Surveillance Center at Case Western Reserve University, who autopsied the brains of McEwan, Whitlock, and the other victim, was quoted as saying that he would not eat venison: "Why should I? I can eat something else. But that's not because I really think there is great danger. I just think the whole issue of prion disease in the United States, both in animals and humans, has to be confronted seriously." Although the brains of the young victims lacked the telltale physical signs common to TSE, five other vCJD patients—all of whom were at least middle-aged and frequently hunted deer or elk—raised further suspicions of a possible link.

Chris Melani, his wife, and their friends, meanwhile, probably became a lot more worried when, in May 2000, By-ron W. Caughey at the National Institutes of Health's Rocky Mountain Laboratories in Hamilton, Montana, demonstrated that chronic wasting disease, at least in the laboratory, is as infectious to human tissue as BSE. Then, in 2002, a cow was experimentally infected with CWD, suggesting one more worrisome similarity between the two diseases. That same year, both the US Food and Drug Administration and the National Institutes of Health announced major new studies to determine CWD's contagiousness to humans and other species.

In 2002 the Wisconsin Department of Natural Resources also announced that three deer shot near the village of Mount Horeb had CWD—the first time the disease had been found east of the Mississippi River. Of the more than 500 deer later shot near Mount Horeb, 15 were infected. Alarmed state wildlife officials began shooting deer from helicopters. Given that hunting is a multi-billion-dollar industry in some states, widespread infection among deer could deliver a severe economic blow. The density of deer in Wisconsin is high, and there is evidence that the higher the density of animals, the more deer become infected. According to Thorne, under some circumstances there is a 100 percent "attack rate," meaning that all the animals will come down with the disease and die from it.

But officials soon realized the futility of trying to eliminate the disease by eliminating deer. By 2010 more than 1,000 Wisconsin deer had tested positive for CWD—and it continues to rapidly spread, with the disease concentrated in two areas of the state. The first is in western Dane and eastern Iowa Counties. The second is located in northern Illinois and extends into southeastern Wisconsin. Continued hunting in the face of the spreading disease means that more diseased deer will be hunted, more harvested deer will go untested, and more hunters will eat infected deer.

In 2005 CWD was documented in a captive white-tailed deer in New York—the state's first. Then it was found in the state for the first time in a wild deer, in Oneida County. The same year, a wild deer in West Virginia tested positive for the disease.

As the need for a better understanding of CWD grew, science lost two of its greatest CWD researchers. In 2004 Tom Thorne and his wife, Beth Williams, died on a snowy late December night when their pickup truck hit a jackknifed tractor trailer on US 287 in northern Colorado. Tom had extensively researched chronic wasting disease in deer and elk and brucellosis in bison and elk. Beth was considered to be among the foremost authorities on CWD.

The disease continued to march east. In 2010 a wild deer in Virginia was diagnosed with CWD. Then came the discovery, in 2012, of CWD in captive deer in Pennsylvania.

Will CWD ultimately spread throughout the suburban East Coast, where deer density is astronomical? The spectacle of blank-eyed, disoriented, and haggard deer wandering aimlessly through backyards in New York and New Jersey would bring home awareness of the omnipresent web of infectious disease as never before. Will this be the disease that

Rutgers University biologist Edmund W. Stiles warned will "knock down" the deer?

Meanwhile, disturbing findings about the nature of the CWD prion continued to emerge. Not only had the disease spread to deer in more than nineteen states; it had also been found in moose for the first time, in Colorado.

Then researchers from University of Wisconsin–Madison discovered that the infectious proteins latch onto certain minerals in the soil and remain infective, revealing yet another possible way for grazing animals to pick up the disease. What's more, binding to certain minerals appeared to actually increase the infectivity. Subsequent research at the University of California, San Francisco, showed that white-tailed deer can shed the prions in their feces long before they show signs of the disease, adding yet another possible route of transmission.

In 2011 the American Dietetic Association published a major study on the possibility of people contracting CWD. "Multiple studies have indicated that a significant species barrier exists that limits the potential of human prion disease resulting from exposure to CWD," the authors wrote. But they warned that "there is concern that increasing exposure to deer or elk meat from CWD-endemic areas may result in infrequent human infections." The concern is based partly on the fact that the prions have been found in the muscle tissue of deer—the part the venison-consumers usually eat. What's more, squirrel monkeys that were experimentally inoculated with brain tissue from infected deer developed the disease. "Primates are capable of becoming infected by CWD prions," the study stated. The researchers also warned that the long incubation time for the disease raises the prospect that some people already infected don't yet know it. As with the victims of mad cow disease, only time will reveal the toll of the new disease.

6.

The infectious particles that cause CWD and mad cow disease had proved themselves capable of spreading in surprising ways. Sooner or later, agricultural officials feared, BSE was going to travel from Europe to cattle in the United States. In an effort to prevent it, as early as 1997 the US Food and Drug Administration had banned the practice of feeding animal by-products to cattle. The FDA stated it was consequently "highly unlikely that a person would contract vCJD today by eating food purchased in the United States." Still, officials worried. For good reason.

In 2003 an adult Holstein cow in Washington State tested positive for BSE. The cow had been imported from Canada two years earlier. In 2005 a cow born in Texas tested positive—the first endemic case of the disease in the United States. The following year a cow in Alabama tested positive.

As efforts to control the disease continued, scientists tried to solve the riddle of BSE's origin. In 2005 a controversial new explanation was published in the *Lancet*. The authors proposed that the disease began when a prion causing human TSE contaminated cattle feed and that the origin was "the Indian subcontinent, from which large amounts of mammalian material were imported during the relevant time period. Human remains are known to be incorporated into meal made locally, and may still be entering exported material." Decades after the emergence, there is still no consensus on the origin of BSE.

Meanwhile, new infections have occurred sporadically in the United States. At the time of this writing, the most recent involved a dairy cow at a rendering plant in California in 2012. The cow had been euthanized after becoming lame and unable to stand; she was then tested and never entered the food chain. Her only offspring was tracked down, euthanized, and also disposed of, although it was not infected.

The California cow had an unusual kind of BSE known as the L-type. Although typical BSE has been linked to contaminated feed, the L-type had not. But extensive research into feeding records showed that the cow had not eaten food from any suspect source. How the cow contracted BSE is unknown. As of 2012, twenty-three cases of BSE had been reported in North America—four in the United States and nineteen in Canada.

7.

Some three decades after the discovery of the disease in cows, memories of BSE still pulse through life at Pitsham Farm. "In retrospect, I'm appalled at what I didn't know about my own cows," Stent confessed several years after the infamous case of cow 142. "I didn't know they were being fed other cows and sheep that had been ground into a powder. We've forced these hoofed grazers into cannibalism. On some farms they're fed growth promotants, and that's probably causing other problems. In many places in the world, livestock is kept in deplorable conditions, all for human convenience and profit. We've put cows on an assembly line and we take them off at the other end and butcher them. Did we really think we could just rearrange the world in any way we pleased? Nobody could have wished for or foreseen this awful thing called BSE. But should we be all that surprised?"

A Chimp Called Amandine: HIV/AIDS

> Going up that river was like traveling back to the earliest
> beginnings of the world, when vegetation rioted on the
> earth and the big trees were kings. An empty stream,
> a great silence, an impenetrable forest . . . the earliest
> beginnings of the world.
>
> —Joseph Conrad, "Heart of Darkness"

1.

It poured before daybreak, and by eight o'clock the smell of raw meat hung in the humidity at the open-air market of Oloumi. As I entered, the morning Air Gabon flight to Paris was passing over Libreville. Inside the market, flanks of duikers, smoked porcupine bodies, and crocodile tails covered the ground. A hairy arm languished in the shadows; a leg was half hidden among brown fur bellies. A monkey lay on its back in eternal slumber, its black, leathery palm half open. My eye fell on an olive-colored tortoise leg, a giant monitor's mottled tail, and the huge, mute beak of a hornbill. Eyes were everywhere, some ratcheted open in blank terror, others squinting as if into sunlight. There was the exhausted, depressed look of a palm-nut vulture and the stunned rosette face of a decapitated mandrill.

A man without a left forearm slid the limp black body of a cat-size colobus monkey off his right shoulder and onto the ground. An aged woman with dried blood on her forearms tossed it onto a pile of intact simian bodies behind her; then she slid another in front of her. Its hair had been singed off. She picked up a knife, carved through soft meat and the stubborn cartilage of a joint, and then loosely wrapped the severed arm and shoulder muscles in plastic and handed it to the man, who tucked it under his good arm and limped away, the casualty of some terrible misfortune.

One by one or in pairs, women with glistening, sunlit faces moved gracefully through the market. By midmorning the tables had been emptied of meat, and by noon Oloumi was quiet but for a few lingering souls packing up their remaining wares. I caught a taxi back to the hotel and ate bright orange carrot and parsley soup for lunch.

I had first come to Gabon in 1994 with colleagues from Tufts University School of Veterinary Medicine, near Boston, to help develop a plan for conserving that country's forests, which French timber companies were rapidly stripping away. Thousands of the loggers living in remote timber camps survived on bushmeat like that featured at Oloumi the morning of my first visit to the country, and the demand was draining the remaining forests of their wildlife, including gorillas, chimpanzees, and nearly every other kind of remotely palatable animal living there. At the time, the threat to wildlife brought about by such hunting was just becoming widely known, thanks to the efforts of the World Wildlife Fund and other groups that pressed for a reduction in logging, arguing that this could in turn reduce the demand for bushmeat.

Few people suspected at the time that the hunting trade might also be contributing to the emergence of AIDS. Whether bushmeat hunting was the vehicle that carried the virus from its source in wild animals to humans is still unproven, but mounting evidence points in that direction. Although the mystery of this elusive, ever changing, and still spreading virus remains, its message has become clear: human health does not exist apart from the larger natural world we share with other species. AIDS is not only a medical issue but also an ecological one.

2.

The conventional medical story of AIDS began in the late 1970s when doctors at three Los Angeles hospitals noted a cluster of illnesses among homosexual men. All of the men had strange life-threatening infections usually limited to people with highly compromised immune systems. In 1981 the Centers for Disease Control and Prevention (CDC) published a report suggesting that a new immunosuppressive disease—apparently caused by a virus—had emerged; the following year the name "acquired immunodeficiency syndrome"—AIDS—was coined. Because the virus attacks the immune system, it leaves the body susceptible to other infections. Many erroneous theories—from nuclear fallout to illicit drugs—arose to explain the malady's emergence.

Over the next few years, records came to light of immune-ravaged patients in Europe a decade or so earlier with symptoms similar to those of AIDS patients. All had a history of close links with Africa, which hinted that the virus may have emerged there. One of the early cases, for example, was that of a Portuguese taxi driver in Paris who had repeated bouts of illness. Several years earlier he had worked as a truck driver in Zaire, where he could have come into contact with prostitutes on the highway between Angola and Mozambique. Belgian physicians had begun to describe patients with similar symptoms in West Central Africa, but they were clueless about the cause.

In the 1990s the theory of the African origin of AIDS was strengthened when researchers in the United States discovered a frozen blood sample that had been taken from a Bantu man in 1959 for malaria research. The man was from Léopoldville, Belgian Congo, near the Congo River—now known as Kinshasa, Democratic Republic of the Congo. When scientists thawed and analyzed the sample, they found human immunodeficiency virus (HIV), the AIDS virus. This Bantu man currently remains the earliest suspected case of HIV-1 infection.

But to know where HIV-1 had come from geographically was not to know *what* it had come from. Was the virus new? If so, what had created it? Had it existed in some unidentified animal reservoir and, given the opportunity, jumped to humans, where it evolved into a deadly new disease? Or had the virus long lurked in humans, but with its spread limited by the isolation of villages where it initially occurred? Had modern transportation and large human migrations suddenly unleashed this long-established infection on the rest of the world?

Understanding the history of the virus, it turned out, was crucial to understanding its future, and while most scientists initially focused their attention on developing treatments for AIDS, a few pursued the lonelier quest to determine its origins. One of those was Beatrice Hahn.

3.

Born and educated in Germany, Hahn trained after medical school in the laboratory of Robert Gallo, the American co-discoverer of HIV. In 1995 she joined the faculty at the University of Alabama, where she worked to pinpoint the elusive origin of HIV-1.

"With the public health calamity of AIDS, it's not surprising that little money and effort was put into discovering the virus's origins early on," she told me. We spoke in the summer of 2002 in her eighth-floor office at the University of Alabama at Birmingham Medical Center in downtown Birmingham. "But that was before people began to realize that understanding the origin of the virus might hold answers to controlling the disease. After that, the interest in discovering the origin of the virus really surged."

What made the quest complicated, Hahn said, is that the AIDS epidemic didn't stem from a onetime event. In fact, two different HIV families have been identified. "Evidence suggests that HIV-1-like viruses, which are responsible for nearly 99 percent of AIDS cases worldwide, have been introduced into humans on at least three occasions," she said. HIV-2, the second human AIDS virus, has jumped from animals to people on at least seven different occasions. HIV-2 is less virulent and is largely limited to parts of West Africa. But both types cause AIDS, and once these viruses entered the human population, they rapidly spread.

Exactly how they jumped is not clear. "We don't know if similar jumps are still occurring, but there's no reason to think they aren't. If anything, the opportunities for jumping have been increasing with the accelerating trade in bushmeat," she said, implying that there may be more HIV types out there that could infect people, if they have not already done so.

Hahn's description of her work brought to mind the image of a contemporary astronomer who tracks asteroids for a living. The chance of an asteroid striking the earth, we are told, is infinitesimal. But given enough asteroids and a long enough time, an impact is inevitable—with potentially catastrophic results. Hahn, however, is not looking at sterile asteroids from outer space. She is tracking evolving microbes from inner space—the space that separates humans from other species. She is tracking the origins of HIV by reconstructing its recent evolutionary tree.

A tall, intense woman with a German accent, Hahn cautioned: "You have to speak of HIV in the plural—HIVs. There is no one 'HIV virus.' There are many."

She interrupted our conversation to take a phone call and then stepped out of the office to have her secretary send a fax. She offered to get coffee, and when she returned a few minutes later, she rested her white mug on the knee of her jeans.

The closest thing to the human immunodeficiency viruses, Hahn explained, are the simian equivalents, SIVs. But "simian immunodeficiency virus" is a huge misnomer. The virus doesn't harm the immune systems of these natural hosts; they don't even seem to get sick. "If we could figure out how simian immune systems deal with the viruses, we might have a clue to controlling AIDS," she said. "That very hope drives my work."

Although SIVs have been known for some time, not until the late 1980s was an SIV discovered that was almost indistinguishable from HIV-2. It came from a sooty mangabey, a small forest monkey of West Africa. That discovery strongly suggested that HIV-1 might also linger somewhere among the hundreds of other simian species in the forests of Africa. But exactly where, no one could say. Its eventual discovery would result from a combination of good science and good luck.

. . . .

4.

In the mid-1980s, the National Institutes of Health sought to develop a vaccine to prevent HIV infection. The research would require chimpanzees as subjects. Chimps used in the study had to be screened to make sure they hadn't previously been inadvertently exposed to HIV through experiments that might have included HIV-contaminated blood. So the NIH began testing more than a hundred chimps held at the Alamogordo Primate Facility at Holloman Air Force Base in New Mexico. More than 99 percent were "clean."

The test results from a chimp named Marilyn, though, were quite different: her blood was full of antibodies to HIV or to something like it. Although the actual virus may or may not have been present in her body then—the test was incapable of showing this—Marilyn's immune system clearly had reacted to something like HIV. Where she had been exposed wasn't clear. Perhaps it was in the African forest of her birth, or perhaps it was during her captivity. It didn't seem to matter at the time; the main concern was that the positive test result had disqualified her from the NIH vaccine experiments. When the scientists who conducted the blood test published the results, though, they also warned keepers and researchers that chimps might carry HIV.

On December 17, 1985, Marilyn died from complications of childbirth. During a routine necropsy, tissue samples were collected and sent to Larry Arthur at the National Institutes of Health in Maryland, who froze them. That was the end of the story of Marilyn, the chimp who might or might not have had HIV. Or so it seemed.

5.

Around 1993, while Marilyn's tissues lay frozen at the National Institutes of Health, a common tragedy was befalling yet another chimpanzee in the forests of West Central Africa. A mother chimp was shot, perhaps in Equatorial Guinea or Cameroon; the location is uncertain. The hunter, or perhaps a middleman, brought the chimp's baby to Libreville and sold her to a childless French dentist and his wife. They named her Amandine. She was given a decorated room of her own in the couple's Libreville home, and in the family's photograph album she can be seen, dressed in toddler's clothes and sunbonnet, accompanying them on a skiing vacation in the Swiss Alps.

Amandine was often ill, in part because her diet lacked the fruits that chimps require. In 1988, her "parents" found Amandine clinging to the swing in her cage, screaming, her body rigid as if paralyzed. They uncurled her arms and lay the chimp prostrate. A physician in Libreville prescribed antibiotics, aspirin, and other medication. Over the next week, the episodes recurred. At one point, Amandine's left hand was clenched firmly closed, and she tried to pry it open with the fingers of her right. Her left leg and arm jerked. She screamed again and eventually went into a series of seizures. In desperation, her caretakers flew her to Franceville, Gabon, to the primate facility Centre International de Recherches Médicales.

Robert Cooper, an American veterinarian, was the center's director of primatology at the time. "When Amandine arrived, she was one sick chimp. She was on the small side for a three-and-a-half-year-old. Her owners kept her in diapers, and she lived the life of anything but a chimpanzee. When she tried to stand up in our clinic, she fell over on the floor and started screaming," Cooper told me recently. A meticulous man, he kept copious notes from the time.

Cooper and a French colleague, veterinarian Jean-Christophe Vié, drew blood, testing of which revealed severe anemia. They gave Amandine a transfusion, and she improved. Vié and Cooper also sent a blood sample for routine HIV testing as part of a program that had been set up to screen wild primates. Cooper and the rest of the staff were stunned when the test came back positive. Unlike the test administered to Marilyn several years earlier, this one identified the actual presence of HIV-1—or, rather, the simian equivalent, which was virtually indistinguishable. What's more, given the unlikelihood that Amandine had been infected by a human, she almost surely had been infected naturally in the wild. This suggested that she—and perhaps her kind—were the natural carriers of HIV-1.

Less than a year later, a second baby chimp with a remarkably similar history was brought to the center for treatment. Her mother had been shot a few days earlier. Two French couples on a weekend outing happened to be passing through the village of Macolamapoye in northeastern Gabon when they came upon this chimp, who had an infected bullet wound in her arm. The couples took the chimp from her captors and made their way to the primate center in France-ville. GAB2, as the chimp was designated, died a few days later, but not before a blood sample was taken. By this time, Cooper had departed—fired, he claims, to clear the way for the Europeans who managed the center to claim ownership of the prized discovery—and Martine Peeters and her hus-band-to-be, Eric Delaporte, had assumed primary responsi-bility for the HIV-chimp study. When they tested blood from GAB2, it came back positive for HIV. Like Amandine, this animal had been infected in the wild. But the researchers' elation at possibly having discovered the origins of HIV-1 was tempered by a sobering thought: the couples who brought in the chimp, who happened to be their good friends, had had their arms and legs badly mauled by the chimp during the trip from Macolamapoye to Franceville. Saliva from the HIV-pos-itive chimp had almost certainly entered their bloodstreams. Would they therefore contract HIV? Delaporte, a physician, was plenty worried. As one might imagine, so were the cou-ples. Delaporte put them on sedatives while they anxiously awaited the results of their own HIV tests.

"The standard practice for accidental blood exposure to HIV was to give a drug called AZT, which was the only treat-ment available, and I recommended that all go on AZT as a precaution," Delaporte told me when I visited him at his office in Montpellier, France. "I got some sent from a physi-cian friend in Paris, and we put them on it. Fortunately, all four of their tests came back negative.

"The fact that our friends didn't get infected doesn't mean that chimps can't transmit the virus through a bite. . . . Maybe it happens only one in a hundred times. We're just thankful it didn't happen in this case."

6.

The fortuitous discoveries of two HIV-positive chimps in Gabon were eureka moments in the quest to discover the origins of the disease, though these two cases were not in themselves definitive. In 1990, however, a third captive chimpanzee in Europe was found to be HIV-positive. Simon Wain-Hobson at the Pasteur Institute in Paris then used detailed genetic analysis to confirm that the virus fragments found in chimps were indeed related to HIV-1. And since their infections had occurred naturally, evidence for chimpanzees as the natural reservoir for HIV-1 was now rapidly mounting.

Then, in early 1998, Larry Arthur at the National Institutes of Health telephoned Beatrice Hahn. "I'm cleaning out my freezer, and I've come across some old tissue samples saved from Marilyn," he told her. "Would you be interested in looking at them?"

"I jumped at the chance," Hahn told me. After analyzing Marilyn's thawed tissues using the latest technology, Hahn and her colleagues found evidence that the chimp had indeed been infected with an HIV-like virus—confirming earlier suspicions. Had Marilyn been artificially infected during experiments while in captivity, or had she acquired HIV in the wild? Hahn set about investigating Marilyn's past, examining the details of every experiment she had ever undergone—and there were dozens—to see if she could have been infected in captivity.

As far as anyone could determine, the chimp, perhaps orphaned by hunters, had been captured as a two- or three-year-old in 1963 and transported to the United States, where she may have spent some time at the Kansas City Zoo. In July of that year, Marilyn was moved to Holloman Air Force Base, which was beginning to collect chimps to use in tests for space flight. After carefully examining Marilyn's history of captivity, Hahn concluded that she had probably picked up her infection in the wild.

Hahn was ecstatic. The finding of yet another infected chimp bolstered her hypothesis. The evidence from Marilyn, Amandine, and GAB2—all belonging to the same subspecies—made a compelling case that the scourge of HIV-1 had originated in *Pan troglodytes troglodytes*.

Yet there was still no direct evidence for the virus in wild-living apes—a critical link in the chain of proof. There was no easy way to study whether these reclusive chimps living in remote African jungles carried the HIV-1 virus—until around 2005, when Hahn and fellow researchers developed a method for identifying evidence of the virus in fecal samples collected from the jungle floor. What's more, they could use the same samples not only to determine the species and sex of the animals but also to identify individual chimpanzees.

Researchers collected almost 600 samples from ten different forest sites in southern Cameroon. The authors concluded that data from the samples "points to chimpanzees in
southeastern and south central Cameroon" as "seeds of the
AIDS pandemic." They concluded that the virus was "probably transmitted locally" before making its way via the Sangha
River and tributaries to the Congo River and then to the
growing city of Kinshasa, where the pandemic began.

Hahn hopes that the findings may one day provide clues
for the control of AIDS. And how is it the chimps apparently
remain healthy in the face of infection? Perhaps the ability
of the chimps' immune systems to neutralize the virus will
shed light on a human prevention. But Hahn's discovery also
led to some larger personal insights. "When I began my work
on the origins of HIV-1, I was a medical researcher," she
said, shifting to a more contemplative tone. "When I learned
that the virus came from chimps and saw how these animals
were being slaughtered, it turned me into an unexpected
conservationist. They are being hunted to the point of extinction, and any clues they may be able to provide could die
with them. It doesn't matter if your goal is to protect public
health or to protect endangered chimpanzees because the
goals are one and the same. We're just not as separate from
the animal world as we would like to believe."

7.

But the discovery of where HIV/AIDS originated didn't
prove how people had first contracted the disease from an-

imals and how it had then evolved and spread to become a staggering pandemic. According to Paul M. Sharp and Beatrice H. Hahn, who published a landmark paper in *Nature* in 2008, the virus probably originally arose between 1900 and 1920. This meant it had been circulating and genetically evolving—mostly in Africa—for some sixty years before it exploded into a pandemic. It may have begun with a few infected individuals in the early 1900s to several thousand by 1960, all in central Africa.

But how did the pandemic emerge in the area of Léopoldville (now called Kinshasa) if the infected chimps lived in the southeastern corner of Cameroon, more than 400 miles away?

As with proof of HIV's origin, the evidence for how it came to infect people was long in coming. Some of the earliest evidence dates from the late 1990s, when Martine Peeters, Eric Delaporte, and their African colleagues began collecting hundreds of meat samples from markets in Cameroon and testing them for viruses. They found that more than 20 percent of the samples were infected with some form of SIV—an extraordinarily high rate of infection and theoretically enough to expose many people to the viruses every day—and they even discovered new SIVs in the process. Which ones are capable of infecting people, and under exactly what conditions they are most apt to do so, remains a mystery.

"When you go to one of these markets, you see a lot of the women with cuts and blood from the animals all over their arms and hands," Peeters said. "In theory, this is a perfect setup for transmission of the virus. Bushmeat is an enormous viral reservoir." Peeters said she believes "there is a good chance there are many more unknown viruses in bushmeat, which is handled by numerous people, from hunters to dealers to the people who take it home. We have no idea what is actually being passed to people through blood and cuts during butchering or from bites from the animals or even their urine."

Peeters is not the only one who fears this unknown. Harold W. Jaffe, an AIDS researcher at the Centers for Disease Control and Prevention, told the *New York Times*, "That is everyone's nightmare, that there is another virus out there that either could be or has been transmitted to humans that we cannot detect with current methods. No one wants to miss detecting the next HIV epidemic."

In 2009 researchers discovered a Cameroonian woman infected with a type of HIV derived from gorillas. Whatever the lack of direct evidence, the broad scientific consensus, according to Hahn, is that "the simplest explanation for how SIV jumped to humans would be through exposure of humans to the blood of chimpanzees butchered locally for bushmeat."

The concern over "transmission events" persists because the continuing bushmeat trade is creating numerous new opportunities for the viruses to jump to humans, who are coming into ever more frequent contact with the blood and other body fluids of the simians that carry them. "The hunting of wildlife, which had always been an important

source of subsistence food in the Congo Basin and throughout sub-Saharan Africa, has increased in the last decades,"
Peeters told me. "Commercial logging operations, many of
them by European-based companies, have led to the building of roads in remote forests. This is followed by massive
human migration and social and economic networks to support the logging. Several thousand people live in many of the
logging concessions, and one of their main sources of food
is bushmeat."

Two of the biggest conundrums of HIV have now been
solved—exactly where it originated and a credible theory of
how it jumped from animals to people. But the question of
timing—why it became a pandemic when it did—persists.
Why in Léopoldville in the twentieth century rather than at
some earlier time during the past millennia?

"The answers may be that, for an AIDS epidemic to get
kick-started, HIV-1 needs to be seeded in a large population centre. But cities of significant size did not exist in
central Africa before 1900," the authors of the 2008 study
concluded. Not only was Léopoldville the largest of the cities, but it was also a likely destination for the virus. The
main transportation routes from the jungles were rivers that
flowed south and drained into the Congo River, right along-
side the booming metropolis.

The emergence, travels, and global explosion of HIV all
speak to the complex ecology of this and many other modern
diseases. It's rarely a simple matter of new viruses or bacteria
jumping from animals to people. It's about the labyrinthine

possibilities for their evolution that human activity and environmental change can provide. People are often unwitting partners in the emergence of new disease rather than passive victims, as we like to believe.

8.

Through the dusk and long into the night I waited on the cement platform at the train station in Lopé, an interior village of Gabon near the Réserve de Faune de la Lopé, a wildlife reserve where chimpanzees and other animals are regularly—and illegally—hunted. It was almost eleven o'clock, and the train to Libreville was already two hours late. Insects the size of hummingbirds buzzed the dim station lights. A freight train passed the platform, the cars fully loaded with tree trunks larger than the fuselage of a Boeing 737. I lay down on the warm cement, with a small suitcase under my head, and dozed. When I awoke twenty minutes later, a woman stood nearby keeping watch over two brightly colored nylon shopping bags with plastic handles. A bevy of flies clustered on the fabric. The woman swatted halfheartedly. When I stood up, I glimpsed the brown fur and black snout of a small animal protruding from the top of one bag. Perhaps this industrious woman was on her way to the market in Libreville to sell her bushmeat.

Many hours late, the passenger train to Libreville finally emerged from the darkness. Over the next six hours, the train frequently shuddered to a stop and then began again on its whining, creaking way. At one point a young woman, assisted by a man, loaded three large sacks, one of them bloodstained burlap, into the coach. By daybreak the train was tunneling through walls of massive green trees and past wide rivers that stretched to the horizon. A light rain melted into forest mist, and the gloom of daybreak gave way to conversation and laughter.

$$\cdots 3 \cdots$$

The Travels of Antibiotic Resistance: *Salmonella* DT104

1.

It was nearly two decades ago that the ominous *Salmonella* DT104 appeared in the eastern United States. On a cold early May morning in 1997, Cynthia Hawley stepped from her clapboard farmhouse in the green-tufted hills of Vermont's Champlain Valley and walked across the drive to feed the calves. To Hawley's dismay, her favorite calf, two-month-old Evita, who'd seemed perfectly healthy the day before, was listless, with sunken eyes and a grotesquely distended belly.

At the time, little was known about DT104. As with most salmonella bacteria, DT104's natural reservoir seemed to be among animals—in this case, cows and chickens. But through undercooked meat or contamination, the bacterium sometimes jumped to people, causing salmonella food poisoning. Although salmonella food poisoning was usually mild or could be cured with antibiotics, DT104 was different, as Hawley would soon discover.

Experienced at treating her animals, Hawley gave Evita a shot of ampicillin and, with a needle, released the gas from her belly. Over the next hour, however, Evita grew worse. Hawley's veterinarian, Milton Robison, arrived and gave Evita fluids for severe dehydration, more ampicillin, and an anti-inflammatory drug, but to no avail. At nine o'clock that night, her sweating head in Hawley's lap, Evita died.

By the next morning several more calves had fallen ill, and Robison returned. He now suspected infection with salmonella, a genus of bacteria that occasionally causes diarrhea in entire herds. An ampicillin injection often solves the problem even when the infection perforates the intestines, invades the bloodstream, and turns the diarrhea bloody. But not this time.

Soon, 22 of the 147 cows at the 600-acre Heyer Hills Farm were ill, including several adults. Within days, 13 had died. Alarmed by the epidemic, Robison sent tissue samples from several of the carcasses to Cornell University's College of Veterinary Medicine in Ithaca, New York, for analysis.

The report from Cornell confirmed Robison's suspicion— but this was no ordinary salmonella. For one thing, analysis showed that the strain was resistant not only to ampicillin but also to four other antibiotics. Concerned that this profile might indicate a new strain, Cornell sent the samples to the National Veterinary Services Laboratories in Ames, Iowa.

Word came back that the strain was *Salmonella* DT104, a deadly variant that for more than three years had been haunting dairy farms—and people—in the United Kingdom and elsewhere. Hawley's sick cows were a prelude to only the third human outbreak of DT104 in the United States—and the first in the Northeast. The first, in 1996, struck nineteen schoolchildren in the small town of Manley, Nebraska, who were infected by drinking chocolate milk contaminated by infected cows. Fortunately, none of the children died. Then, less than six months before Hawley's cows got ill, there was an outbreak in Yakima County, Washington, and in northern California, which sickened more than 150 people who had eaten unpasteurized Mexican-style soft cheese. In all her years working with dairy cattle, Hawley had certainly never seen or experienced anything like it. And she herself had not become sick—yet.

The bacteria existed not only on farms but also on su-permarket shelves—at least in the Washington, DC, area, where in 2001 the US Food and Drug Administration col-lected 200 samples of chicken, beef, turkey, and pork from three different stores. One in five of the samples was con-taminated with one form or another of salmonella—includ-ing DT104. The FDA also identified a bacterial strain resis-tant to twelve different antibiotics. Eaten undercooked, any of those samples could have given consumers food poison-ing. What made DT104 dangerously different?

The particular part or segment of a bacterium that makes it drug-resistant can occur in different places on the microbe. Usually a tiny segment on the bacterium neutralizes the antibiotic by either breaking it down or preventing the bacterium from ingesting it. These tiny bacterial segments can be shed or acquired as environmental conditions require. That is why suspending the use of certain antibiotics can cause resistant bacteria to slowly lose their resistance—and why prudent use of the drugs can possibly restore their curative powers.

But DT104 carried its resistance in a more or less permanent form—that is, within the cell's genetic material. The chance of its shedding the resistance, even in the absence of antibiotics, was reduced. Once DT104 became resistant to a drug—and the DT104 that struck Hawley's farm had already become resistant to at least five of them—those drugs would probably remain impotent against the strain. The bacteria had mutated, and as a consequence resistance had become essentially permanent.

2.

The scientific study of salmonella began in 1885, when pathologist Theobald Smith isolated the organism in tissue from pigs. But the bacterium was named after Smith's supervisor, veterinarian Daniel E. Salmon of Cornell University, who had brazenly usurped credit for the discovery.

The term "salmonella" refers to a genus of bacteria that lives in the intestines of many species. In many animals it doesn't cause illness, but in others—including humans—the sickness can be fatal. This depends in large part on the kind of salmonella involved. Salmonella and its different forms often cause food poisoning.

Some types of salmonella, including those most dangerous to humans, emerged through our own practices, including the widespread use of antibiotics. Although some bacterial strains may resist antibiotics naturally, most become resistant because of overuse of the drugs. Through the phenomenon known as natural selection, bacteria frequently challenged by these drugs grow accustomed, in a sense, to their presence. Just as every human is different, so are most individual bacteria. And just as some people seem more resistant than others to a particular illness, so are some bacteria better equipped than others to survive in certain environments. When the immediate environment is awash in antibiotics, the vast majority of targeted bacteria may die, but some of the better-suited ones will survive. This is the first step toward antibiotic-resistant disease.

The surviving bacteria form the basis for the next generation and, of course, pass their traits to their offspring. Hence, more bacteria in the next generation will survive exposure to the same antibiotic. Continual exposure to drugs can make each generation more resistant.

One environment where bacteria are frequently flooded by antibiotics is in large livestock operations, where producers frequently treat their cows and other animals with drugs to prevent epidemics in the unsanitary and overcrowded conditions commonplace in the industry. In the short term, it's cheaper to keep animals drugged than to keep them clean. Animals fed a steady diet of antibiotics with their grain also grow a little faster, thereby making the producers extra money. In addition, farmers often feed antibiotics to newborn calves—again, for the sake of short-term efficiency. These producers want to put the mother back in the milking parlor shortly after birthing, so they immediately send the newborns off to join thousands of other calves at a grow-out facility. Deprived of the natural antibodies in mother's milk, the newborns are given antibiotics instead.

Unfortunately, individual bacteria in the cows' intestines that survive this onslaught of antibiotics—and almost always some do—are resistant to antibiotics. Any number of genetic quirks, such as the particular makeup of the cell wall, for example, may let one bacterium survive where another would die.

The final step in human salmonella infection occurs when these resistant bacteria infect a person through undercooked meat or other contamination—such as on Heyer Hills Farm. Residual bacteria, which may include salmonella, often linger on the meat we bring home from the supermarket. Before cooking, when we open the package or prepare the meat, juices can contaminate the kitchen countertop and salad spinner, making it into refrigerated food and eventually into our mouths. Once taken into the stomach, the bacteria may pass into the intestines and enter the cells lining the intestinal wall, causing inflammation and pain and possibly fever. In most healthy individuals, immune cells in the intestines will kill the invaders, and the illness may pass as transient diarrhea. In severe cases, as in older persons, infants, or others with weak immune systems, or if the bacteria are especially virulent, the invaders perforate the intestines, causing bloody diarrhea and then escaping through the intestinal wall into the bloodstream. If a bacterium happens to be a virulent one causing severe food poisoning, such as *Salmonella* DT104, a short course of antibiotics usually cures the infection. But if the offending bacterial strain has already adapted to antibiotics in the farmyard, it could survive the treatment. A normally curable case of food poisoning could rapidly become fatal.

When a strain of bacteria is subjected long enough to a variety of antibiotics, very powerful offspring with a wide range of resistance are likely to develop. To make matters worse, a bacterium that grew resistant while living in an antibiotic-laden fish-farming operation, for example, may be picked up and carried by a bird to a stockyard, where that bacterium might trade parts with another and pass on its resistance in the process. Because the highly adaptive *Salmonella* DT104 can infect birds, cattle, amphibians, and many other species, antibiotic-resistant illness can quickly spread across species. Not surprisingly, some of the most dangerous forms of salmonella have evolved in large-scale livestock operations. These epidemics often have begun with large producers and later infected smaller farms, such as Heyer Hills.

The incentive for farmers to use antibiotics is almost irresistible because these drugs help animals grow faster and produce more. The first approved use of antibiotics in livestock dates back to 1951, when the FDA began to allow them as animal feed additives to help pigs, chickens, and livestock gain weight more quickly. As an advertisement from American Cyanamid said, "Why wait until disease has caused weight losses, poor egg production, feed waste, . . . and dead birds? Feed AUREOMYCIN Chlortetracycline to chickens and turkeys *continuously at* HIGH LEVELS and prevent these losses! Give them *internal* sanitation. . . . Heavier, top-quality meat birds! . . . And PROFITS . . . *several times higher!*"

A half century ago, resistance that did develop was probably slow to travel elsewhere. Today, however, bacteria travel the microbiological equivalent of the interstate highways— or, rather, international air routes and shipping lanes. Where bacteria are concerned, Heyer Hills Farm could be just a shipment of cows or feed away from the United Kingdom. And the meat on the shelves of a supermarket in Washington, DC, could have been in California or Texas the day before. In a world of incessant global commerce, there is no one to whom we are not ultimately connected.

3.

The outbreak of DT104 at Heyer Hill was not the first drug-resistant form of *Salmonella* DT104 to emerge from livestock agriculture. Rather, the first major documented epidemic struck the United Kingdom in the early 1960s. So-called type 29 defied antibiotics on a scale unprecedented at the time. As tracked by London's Public Health Laboratory

Service, by 1963 type 29 had become resistant to two antibiotics commonly used in livestock. It soon picked up resistance to tetracycline and, later, to two more antibiotics. By
1965 more than 95 percent of *Salmonella* type 29 tested in
cattle in the United Kingdom showed antibiotic resistance,
and some rare forms had armed themselves against seven
antibiotics. Of some five hundred confirmed human cases
of *Salmonella* type 29 food poisoning in the United Kingdom, six were fatal. The epidemic's origin was traced to a
livestock dealer who, despite heavy use of antibiotics, had
sick calves—and was selling them throughout the United
Kingdom. After being charged by the government with illegal sale of sick calves, the dealer, a Mr. Atkinson, apparently
committed suicide by slamming his car into a tree.

In the mid-1960s, E. S. Anderson of the Public Health
Laboratory Service concluded that the type 29 outbreaks
were "almost entirely of bovine origin" and warned that "the
time has clearly come for a re-examination of the whole
question of the use of antibiotics and other drugs in the
rearing of livestock." An editorial published around that
time in the British magazine *New Scientist* argued that the
use of antibiotics to make animals grow faster "should be
abolished altogether."

With physicians fearing the same could happen in the
United States, a 1968 editorial in the *New England Journal
of Medicine* warned that antibiotics could become useless,
sweeping away a major modern line of defense against infectious illness. Even common and treatable illnesses such

as pneumonia could produce vast epidemics with numerous deaths. "Unless drastic measures are taken," the article said, "physicians may find themselves back in the pre-antibiotic Middle Ages in the treatment of infectious diseases."

By the early 1970s the Food and Drug Administration agreed that there was cause for alarm and said that "there is ample data now in the literature to support more rigid control of antibiotics in animal feed and water." As C. D. Van Houweling, a veterinarian and chairman of the FDA's task force on antibiotics, explained, indiscriminate antibiotic use "favors the selection and development of single- and multiple-antibiotic-resistant bacteria . . . and could produce human infection." The logic applied to many kinds of bacteria. The FDA concluded in the early 1970s that, at the least, licenses for use of antibiotics as growth promotants should be revoked.

The drug industry immediately launched a counteroffensive in the media and even in scientific journals. In a 1973 article published in *Advances in Applied Microbiology*, Thomas H. Jukes, a former biochemist for Lederle, one of the first commercial producers of antibiotics for livestock, blamed the FDA's conclusion partly on "a cult of food quackery whose high priests have moved into the intellectual vacuum caused by rejection of established values." He cited as evidence of the cult two bills then before the United States Congress that would "authorize definitions for 'organically grown food which has not been treated with preservatives, hormones, antibiotics or synthetic additives of any kind.'"

Jukes also advocated that antibiotics be routinely used in some human food. "I hoped that what chlorotetracycline did for farm animals it might do for children," he wrote, referring to less disease, fewer illnesses and deaths, and "slight to moderate increases in growth." Antibiotics, he believed, could compensate for malnourishment and for the overcrowded and often unhygienic living conditions of many people in the developing world, just as they did for cattle. "This sounds like the conditions under which chickens and pigs are reared intensively," he wrote, concluding that similar benefits could result for humans.

Faced with vehement industry opposition, the United States took no action to limit antibiotic use in livestock at the time, but in 1970 the British Parliament did ban the use of almost all antibiotics to promote growth—despite similar opposition from livestock producers and the drug industry in the United Kingdom. Over the next six years, the incidence of *Salmonella* type 29 declined in the United Kingdom, presumably because the reduction in antibiotic use permitted populations of nonresistant forms of the bacteria to build up again. In other words, in the absence of antibiotics, resistant bacteria had no survival advantage over nonresistant ones. Unfortunately, the legislation did not restrict new antibiotics that would soon come on the market. A decade later, use of these new antibiotics in livestock would precipitate another salmonella epidemic—a second warning that continued routine use of large amounts of antibiotics posed a grave public health threat.

From about 1973 to about 1980, completely new types of drug-resistant salmonella began to appear among cattle in the United Kingdom—including a new multiple-drug-resistant strain that struck more than fifty farms in southern England and spread to Cambridgeshire and Yorkshire. Years later, the folly of the situation was summed up in testimony to a House of Lords committee investigating antibiotic use. The way antibiotics were being used, the witness said, reminded him of "the man who threw himself out of the Empire State Building and as he passed each window he said, 'So far so good, so far so good!'"

The National Office of Animal Health (NOAH), the representative of animal drug manufacturers in the United Kingdom, argued that there was no problem with continued routine use of antibiotics in livestock because "new antibiotics are being developed all the time." As a resistant bacterial strain developed, so the argument went, the industry would develop a new drug to counter it. NOAH did not point out that the "new" antibiotics were mostly spin-offs of existing ones, and therefore the bacteria would very likely be as resistant to them as to their immediate antecedents. The discovery and development of new antimicrobial drugs for multi-resistant organisms would, in fact, soon begin to slow as companies shifted their dollars away from research on new antibiotics and toward pharmaceuticals that carried a higher profit margin, such as cancer drugs.

In response to continuing salmonella epidemics in the United Kingdom and the United States, in 1977 the FDA finally proposed banning the use of penicillin and tetracycline as growth promotants unless the pharmaceutical industry could show that this use was safe. Although no such evidence was forthcoming, the FDA failed to withdraw approval for the drugs. No action would be taken on this issue for another thirty-five years.

Not long after the 1977 proposal, the National Academy of Sciences' National Research Council (NRC) completed a report on agricultural antibiotics. By that time, no fewer than half a dozen weighty scientific evaluations had already been completed, with the broad consensus in the primary scientific literature that overuse of antibiotics in agriculture posed a health risk to humans. The NRC report, however, concluded that there was not *absolute* proof—no smoking gun. (Then again, the NRC pointed out that it was virtually impossible to definitively link antibiotic use in animals to food poisoning in humans caused by drug-resistant bacteria. This is because evidence of transmission is the contaminated food itself, which has usually been disposed of before people get sick and an investigation is begun.) In the end, the report, which was written by a committee chaired by Raoul Stallones of the University of Texas School of Public Health, a paid consultant to several animal-drug companies and an outspoken advocate for unlimited antibiotic use in livestock, argued that the practice should be continued for its economic benefits. Sometime after release of the NRC's report, Stallones wrote, "If the decision were mine, the hog farmers could use all the antibiotic drugs they wish to make the pigs grow."

Far from presenting any evidence that the use of antibiotics in livestock feed was safe, study after study proved its risks. In 1982 a smoking gun was unexpectedly found when researchers at Harvard Medical School traced tetracycline-resistant illness in humans to tetracycline use in animals. By using genetic fingerprinting to exactly match the bacteria in the livestock with the bacteria in the patients, Thomas O'Brien and his colleagues had, in effect, solved the case without using food as the witness. The FDA considered the study definitive, concluding that the issue "certainly has been studied sufficiently" and that no further evidence was needed to justify limiting or banning the use of certain antibiotics in livestock. Van Houweling, who had argued in favor of the earlier FDA ban, meanwhile had become a consultant to the hog industry. In response to the FDA's latest conclusions, he made an about-face, stating that "history has shown that it doesn't make that much difference" if the drugs are banned in feed. (Britain's ban of certain antibiotics in 1970, of course, had suggested exactly the opposite.)

In the face of continued congressional opposition to limiting antibiotic use in livestock, the FDA still took no action on conclusion of harm. The agency's budget, after all, was in the hands of the same appropriations subcommittee that handled the budget of the US Department of Agriculture, which was heavily influenced by agricultural interests.

Throughout the 1990s, scientific evidence continued to mount. In 1997, the year that Cynthia Hawley grew ill, the World Health Organization reinforced recommendations that had actually been made three years before: "The use of any antimicrobial agent for growth promotion in animals should be terminated if it is used in human therapeutics; or known to select for cross-resistance to antimicrobials used in human medicine."

4.

On Friday, May 16, Cynthia Hawley drove to Burlington, about twenty miles away. While waiting to see her hairdresser, she felt a sharp abdominal pain, severe enough to make her double over. She lay down in a back room while her hairdresser phoned Hawley's sister and mother, who drove to Burlington to pick her up. Once she got home, the diarrhea and vomiting began. Over the course of that night, she grew weaker. The next day, with Hawley unable to retain any fluids at all, her mother insisted on driving her to Northwestern Medical Center in nearby St. Albans. By the time they arrived, she was unable to walk unassisted.

"What brings you here?" the attending physician asked when Hawley finally reached the emergency room.

"Acute gastroenteritis," she groaned, accustomed to using medical terminology on the farm. Remembering the veterinarian's warning that the infection was highly contagious to humans, she added: "The cows have it. It may be DT104."

That physician had never heard of DT104. The following day, the case was assumed by Mara Vijups, a physician trained at the University of Vermont, who'd never heard of it either.

"Cynthia was medically off the charts," Vijups told me when I visited her clinic in Vermont. "I'd never seen anyone that sick from salmonella. Her blood count showed her blood was very toxic. Her face was gray. She was losing massive amounts of fluid through bloody diarrhea. It was all we could do to keep her alive that first night. She hadn't eaten in days; I really feared we'd lose her. We pulled up some articles on DT104, and I called Cynthia's veterinarian and the state veterinarian, since they had experience with it in animals. They warned me of what to look out for. I was very scared."

Vijups was used to having several drugs at her disposal to treat patients with severe salmonella infections. Ampicillin almost always worked, but as Vijups knew from her crash course in DT104, that drug would be powerless in this case. A combination antibiotic known as Bactrim was another option, but Hawley, like many people, was allergic to it. That left Vijups to ponder two remaining life-or-death options for her patient. One was cephalosporin. Although the drug was often prescribed for salmonella food poisoning, its effectiveness against the infection had not been widely studied, leaving open the possibility of unexpected failure in the face of DT104. The second option was fluoroquinolone, which had a long and distinguished track record against more traditional forms of *Salmonella* DT104. But DT104 had begun showing signs of fluoroquinolone resistance in the United States and Europe. Still, it was the best hope, and Vijups decided to prescribe it—and pray for the best.

By Wednesday, May 21, Cynthia Hawley had emerged from her stupor long enough to catch the *CBS Evening News* with Dan Rather, who happened to be reporting on another deadly DT104 outbreak in England. The next morning, when Vijups came into her hospital room to report the US Department of Agriculture's test results on her bacterial culture, Hawley interrupted: "It's definitely 104. I saw the news last night. Once the cows' diarrhea started getting watery and bloody, they were dead."

"Yeah, DT104 is what we're dealing with," Vijups confirmed. "The good news is that it's sensitive to fluoroquinolone, the drug you're on."

Although she didn't tell Hawley at the time, Vijups had a deep personal response to Hawley's illness: Vijups's own grandmother had died from salmonella food poisoning decades before. "The stories my mother told about my grandmother's death were always with me when I was treating Cynthia," she said. "I was haunted by the picture of what the world must have been like with no antibiotics to treat the illness. As a physician who encounters treatment failures because of antibiotic resistance, I have moments of fear that we're moving back to that time when infections, even mild ones by current standards, will become fatal again."

5.

From the very first documented human outbreak of DT104 in Great Britain in the 1970s—when it struck seven people in Airdrie, Scotland, including five from one family—the traits of the new bacteria were frightening. DT104 killed more cows and made people much sicker than had been the case in early *Salmonella* DT104 epidemics, and its embedded resistance set it ominously apart. But there was another aspect that made it unusual. Antibiotics commonly develop resistance "locally"—that is, their repeated use in one setting can confer resistance to the bacteria involved in that particular infection. While those bacteria themselves could spread from one person to the next, taking their resistance with them, in some cases resistance can spread from one type of bacterium to another. For example, if fish-growers treat a fish disease with antibiotics, that fish illness can become resistant. The gene or genetic material that causes that resistance can then spread to a human pathogen, such as salmonella. That, it would turn out, was the case with DT104. Although the drug-resistant bacterium was first detected in the United Kingdom, its resistance seems to have come from far away.

No one understands this better than Frederick J. Angulo of the Centers for Disease Control and Prevention, who formerly headed the CDC's National Antimicrobial Resistance Monitoring System (NARMS) and now is chief of the Global Disease Detection Branch at the CDC. As head of NARMS, it was his job to identify and track dangerous characters of the microbial world such as DT104. Angulo is a cheery middle-aged man with a serious focus when it comes to bacteria. "In 1996 a CDC colleague in Geneva, Switzerland, e-mailed me an article about a bacterium that was isolated from a kitten in England," he told me when I visited his office in June 2002. "We didn't know at the time it had already reached the US. The appearance of DT104 was remarkable. It didn't slowly move from one country to the next, leaving a trail of intermediate forms as it evolved. The complete bacteria just exploded globally all at once, including in the United Kingdom and western Europe, Japan, and other countries.

"The first documented human infection in the US, it turns out, had been a man from Kansas in 1985. However, we did not become aware of the problem until the mid-1990s. The case in Kansas suggests that the bacteria had been lurking around, perhaps in cattle, in this country. It just took the bug a while to cross over to people through food. I originally thought DT104 had emerged in livestock agriculture, and that may still be the case, but by 2000 there was growing evidence that the resistance package of DT104 came not from livestock but from farms of a different sort—fish farms."

Two of the resistant genes in DT104 have been traced back to fish bacteria that commonly occur in aquaculture. Both of the genes that code for tetracycline resistance first occurred in a bacterium causing disease in farmed fish. The DT104 gene that codes for resistance to chloramphenicol is almost identical to a resistance factor that comes from another bacterium common to fish raised in captivity.

"Three of DT104's most distinguishing traits are directly related to its resistance," Angulo said. "Two of the particular resistances the salmonella carries are quite rare. In some cases, the only other place that occurred, and that was a few years before the appearance of DT104, was in bacteria that lived on farmed fish in Southeast Asia."

"How," I asked, "would the packet of resistance genes have moved from fish bacteria to the *Salmonella* DT104 outbreaks years later?" Angulo didn't know for certain, but he suggested several plausible scenarios.

According to Angulo, a *Salmonella* DT104 bacterium probably would have had to encounter a fish bacterium and pick up its resistance, perhaps in a pond filled with waste from both cattle and fish farms. Or a bird infected with the microbe could have visited an aquaculture facility and defecated in the water. "Wherever it happened, once salmonella had picked up the resistance, it could have gotten into fish meal made from discarded fish products. Fish meal is a common supplement in cattle feed. Contaminated feed could have rapidly been shipped to Europe, Japan, and the United States, where it then infected cattle in those countries. Then it was just a matter of time before it jumped to people."

The idea of bacteria being spread around the world in animal feed is not far-fetched. In the 1970s, a rare type of salmonella that caused international outbreaks was traced to fish meal from Peru. Fishermen had dried the fish on the decks of their ships, and seabirds infected with the salmonella had defecated on the fish during the drying process. The bacteria became part and parcel of the fish meal, which was quickly spread through international trade. The fish meal was fed to poultry. Undercooked, the meat then infected people. Although the bacteria from Peru were not resistant to antibiotics, the incident showed how quickly bacteria could spread through trade. "There are other plausible scenarios on how DT104 got here, such as by dissemination via breeding stock," Angulo said, "but dissemination via fish meal is one way."

"The main point," Angulo concluded, "is that DT104 is a complex story of animals, their diets, food production, and global commerce. The story has many interlocking pieces, but it comes down to people impacting global systems and disrupting the natural ecology of animals through artificial diets and intense husbandry. This, in turn, impacts our health."

6.

Cynthia Hawley had been lucky: fluoroquinolone worked. She was one of millions of beneficiaries of a drug that, when it came on the market in the 1980s after more than thirty years in development, was immediately hailed as a breakthrough treatment for many infections, including the severest cases of salmonella food poisoning. In retrospect, given the drug's unique value in treating potentially fatal human disease and all that was known at the time about antibiotic resistance, it is hard to fathom why several European countries quickly approved the lifesaving antibiotic to prevent outbreaks of diarrhea in calves and respiratory disease in overcrowded poultry—setting off a new round of antibiotic resistance that would put people in even worse shape than before.

The Netherlands approved the use of fluoroquinolones in livestock in 1987. Soon afterward, the bacteria responsible for a major type of human food poisoning began to show resistance to fluoroquinolones. Six years later, in 1993, despite studies documenting the risks, Denmark licensed them for veterinary use. Soon, the first signs of human resistance to the drug were documented in that country. In 1998, five people associated with a Danish swine slaughterhouse were stricken with a fluoroquinolone-resistant strain of DT104. The bacteria had quickly spread from the pigs to the slaughterhouse workers, who in turn infected nurses at the hospital. The bacteria also contaminated some meat products, infecting a woman who tasted a raw meatball before frying it. In the end, more than twenty additional people fell ill, eleven were hospitalized, and two died.

Also in 1993—twenty-five years after one government committee in the United Kingdom warned of the growing risk of antibiotic resistance in livestock—another part of the government licensed fluoroquinolone for treating and preventing illness in turkeys and chickens. Two years later, 16 percent of *Salmonella* DT104 cultured from farms in the United Kingdom showed some resistance to the drug, and by 1996, fluoroquinolone-resistant salmonella infections were sickening people.

Given the rapid development of fluoroquinolone resistance in the Netherlands, Denmark, and the United Kingdom following its use in agriculture, the FDA seemed to have an airtight case for rejecting the drug manufacturer's application, in 1995, to sell fluoroquinolones for use in poultry in the United States. On this subject the FDA also had the full support of the CDC, which sent nine letters to the agency urging it to reject the application. Nevertheless, in 1995 the FDA granted approval for use of the cutting-edge antibiotic to treat respiratory disease in poultry. By 1997 salmonella in the United States had begun to show resistance to fluoro-quinolones. By 2000, 1.4 percent of salmonella infections showed some resistance, with the percentage quickly rising.

This was strong evidence, according to the FDA, that the use of fluoroquinolones in poultry posed a risk to human health. In 2005—a decade after approving the use of fluoroquinolones for poultry—the FDA reversed itself and withdrew approval. It had actually proposed the ban five years earlier, prompting the Bayer Corporation to launch a five-year battle against it. The director of government and industry relations for Bayer, veterinarian Dennis Copeland, insisted, "The consensus is that there is no public health risk." (Abbott Laboratories, one of the two manufacturers of fluoroquinolones for poultry, had withdrawn its product before the FDA officially proposed the ban.)

Alexander S. Mathews, president and chief executive officer of the Animal Health Institute in Washington, DC, which represents manufacturers of pharmaceuticals, vaccines, and feed additives, claimed that "there is no scientific evidence that salmonella food poisoning has been linked to farm use of antibiotics." Richard Carnevale, also of the Animal Health Institute, declared, "There is no clear documentation that use of antibiotics in these animals was responsible for the emergence of the multi-drug-resistant strain of salmonella."

Patrick Pilkington, vice president of Live Production Services at Tyson Foods, stated that "scientific information currently available shows no conclusive evidence of a connection between the veterinary use of fluoroquinolones and antibiotic resistance in humans."

In 2001 Richard L. Lobb, a spokesperson for the National Chicken Council, told a reporter for the *Village Voice* that fluoroquinolone actually "improves the gut health of the bird and its conversion of feed. . . . And if we are what we eat, we're healthier if they're healthier." Of course, the birds themselves, often deformed or weakened by their artificially rapid growth resulting from unnatural feed, their cramped and unsanitary quarters, and their water supply, which is sometimes laced with antibiotics, were profoundly unhealthy. And the spokesperson also neglected to mention that by eating the birds, people risked ingesting antibiotic-resistant bacteria.

Meanwhile, scientific warnings from the CDC, the FDA, and the American Medical Association largely echoed the warnings from the 1960s that the use of antibiotics for promoting growth in farm animals should be banned. Knowledge had marched on even as common sense stood still.

Antibiotics used on the farm not only make livestock-associated bacteria resistant to some antibiotics but also can remain active after passing through the animals. The drugs then end up in bacteria-rich waste lagoons, and this medicated sludge is often spread on croplands as fertilizer, where the antibiotics and drug-resistant bacteria enter groundwater or surface water and then infiltrate the soil. The CDC has found significant levels of three different antibiotics in lagoon wastewater drained from industrial feedlots, agricultural drainage wells, and associated water sources. This wastewater contaminates streams, rivers and aquifers, and lakes and their shores, exposing those who swim there or eat fish from the seemingly pristine waters.

Fluoroquinolones have also been detected in wastewater treatment plants in Europe. One study found high levels of antibiotics, very likely from nearby cattle operations, in two lakes in Switzerland. And researchers have reported the presence of antibiotics in river water and sediments in Italy. Other antibiotics have been detected in sediments under fish farms. This is not surprising, given that up to 80 percent of the antibiotics used in aquaculture ends up in the environment.

7.

On May 26, 1997, Cynthia Hawley left Northwestern Medical Center and returned home to Heyer Hills Farm. Years later, as we sat in the farmhouse kitchen on a blistering July afternoon, she lamented the globalization of world trade or whatever it was that had permitted DT104 to be visited upon her farm and family. Outside the kitchen window, whose sill was adorned with a plaster cast of a black-and-white Holstein, a row of evergreens stood guard along the edge of the north pasture.

Hawley was stunned by Frederick Angulo's theory that parts of the bacteria that struck her farm could have come from as far away as Asia. The notion that pieces of a bacterium could hop from fish in Thailand to, perhaps, a bird and then reassemble themselves as they travel across continents, only to strike at the very heart of her family's health and income, was a sobering reminder of the dangerous complexities of modern life. She nodded as if to acknowledge not only the logic of this scenario but also its inevitability. Species intermingle all the time. Humans are connected not only to one another but also to the myriad other species, seen and unseen, with which we share the earth.

"Our family farm, with about 200 head of cattle and 600 acres, used to be considered really huge in this area," Hawley said. "We were a big fish in a small pond; now we're a small fish in a big pond of corporate agriculture. You have to get bigger to survive because you need to produce quantities in order to compete. 'Farm' is becoming a misnomer. It's pretty much industry now. I do not like to see what's happening to the animals because of it."

Later, we walked out to the garden, which Cynthia's mother, Marjorie Heyer, was tending. "The more intensive farming gets, the more props you need," Heyer said. "You crowd the animals to save every cent you can on space; then you have to give them more antibiotics to keep 'em healthy. I'm not saying that's where DT104 came from. I'm just saying that forty years ago what we worried about was nutrition and how to feed the cows right. Now it seems like there is a lot more we have to deal with and worry about, especially after this DT104 thing. What's next?"

Following her parents' inspiration, when Cynthia was twenty-nine she had married a farmer, Brian Hawley, who owned a large independent dairy operation nearby. Even then, it wasn't as if the couple had a secure hold on their dream. "One January morning my husband went out to start a tractor. The tractor ran over him, and he was killed," she said, her stare fixed on a distant memory. Determined not to let their vision die with him, Cynthia operated the large farm on her own, with hired labor, for the next twelve years. In the fall of 1996 she sold the farm and moved back to Heyer Hills Farm, where she had grown up. It was not exactly as if she had come full circle; she had traveled down a river. Life had always been a river, but now, fed by new tributaries from all parts of the world, the river of life at Heyer Hills Farm felt swifter, more dangerous, and less predictable than ever before.

8.

Early fears that the resistance of DT104 would become permanent were soon realized. Five years after Hawley's encounter, fifty-nine laboratory-confirmed cases were identified in nine states. The bacterium was still resistant to several important antibiotics. Almost half of the victims were hospitalized for an average of four days. The illness was traced back to ground beef that the patients had bought from grocery stores. Then, in 2009, DT104 struck in Colorado, Kansas, Missouri, Nebraska, New Mexico, Utah, and Wyoming. The outbreak led to the recall of nearly 500,000 pounds of ground beef at a meat-processing plant in Denver.

By then, if the FDA seemed to have stood still on regu-lating many antibiotics, it was at least moving ahead with cephalosporins, an important new type of antibiotic used to treat pneumonia, strep throat, and many other infections in people but one that accounted for less than 25 percent of all antibiotics used in livestock. In 2008 the FDA proposed banning cephalosporins in response to the evidence showing that use of the drugs in livestock was threatening their effec-tiveness in treating serious salmonella infections, including those in children. But the FDA later withdrew the proposal, saying it needed more time to study the issue.

And after thirty-five years, the FDA had still failed to withdraw approval for penicillin and tetracycline, as required by its 1977 findings that their use in livestock was a public health menace. But if the FDA had forgotten, many public interest groups had not, including the Union of Concerned Scientists, the Center for Science in the Public Interest, and the Natural Resources Defense Council. In May 2011 the NRDC, joined by other advocacy groups, filed a lawsuit against the FDA for failing to act on the 1977 findings—that is, failing to withdraw approval for use of the two antibiotics in promoting the growth of livestock. In response to the law-suit, in 2011 the FDA rescinded the 1977 documents con-taining its findings, but it did not overturn the findings them-selves. The courts pointed this out and determined that the FDA was still obligated to follow through. Unless the agency could show that the use of these antibiotics posed no risk to human health, they would have to be banned. In either case, the agency could not simply pretend the 1977 findings did not exist—even after thirty-five years of ignoring them.

While the use of penicillin and tetracycline as feed additives in livestock remains in a legal tangle, the FDA recently moved forward on the use of cephalosporins after their false start in 2008. In 2012 the agency banned their extra-label use. Although the drug was a minor part of the veterinary armamentarium for livestock anyway, the FDA rule was a small step forward.

The FDA also put in place "guidance" on the judicious use of other antibiotics in livestock. Unfortunately, this guidance was not binding. What's more, it actually endorsed the principle of using drugs to prevent disease in livestock—a loophole through which the drug makers could drive delivery trucks. After all, disease prevention could mean feeding small amounts of the drugs to livestock throughout their lives—the very kind of low-dose use that allows many more bacteria to survive. This, in turn, can quickly select for bacterial resistance to the antibiotic.

Even in the face of weak FDA guidance, the American Farm Bureau Federation claimed that there was insufficient evidence of "on-farm antibiotic use that demonstrates a meaningful risk to humans." The Farm Bureau went so far as to claim that scientists from the CDC, the National Institutes of Health, and the US Department of Agriculture had stated that "there is no scientific study linking antibiotic use in food animal production with antibiotic resistance."

Even if the recent FDA action leads to decreased use of antibiotics in the United States, antibiotic resistance, like many infectious diseases, will remain a growing global problem. Currently China is both the largest producer and the largest consumer of antibiotics in the world, according to the National Academy of Sciences. Nearly half are used in animals. In 2012 a team of American and Chinese scientists genetically analyzed samples of manure from Chinese hog operations and found 149 resistance genes. These included resistance genes for most major classes of antibiotics.

Stuart B. Levy, MD, director of the Center for Adaptation Genetics and Drug Resistance at Tufts University, worries that the problem of antibiotic resistance has become ubiquitous. "If antibiotic-resistant bacteria in the environment glowed red, we would see them everywhere," he said. We would see them on lawns and in sinks and toilets, along rivers and lakes, in forests and along coastlines. We would see them from Alaska to Arizona and, now, from the farmlands of California to the pig farms of China.

Of Old Growth and Arthritis: Lyme Disease

1.

About the year 1700, John Harrison of Long Island, New York, bought from the Lenni-Lenape Indians a 17,000-acre tract of oak-hickory forest near what is today the city of New Brunswick, New Jersey. At the time, Harrison's purchase was but a tiny grove within 100 million acres of woodland that stretched from Virginia to New England and west to the Mississippi River.

In 1701 Harrison sold 10,000 acres of his land to a group of Dutchmen, who divided it into eight parcels. South Middlebush, the first road through the region, crossed the tract north to south, subdividing the eight parcels into sixteen, and other subdivisions began to be made. Cornelius Wyckoff, one of the buyers of Harrison's land, for example, gave 300 acres to each of his four sons, who built houses and

. . . .

cleared land for crops and livestock, leaving behind several forested woodlots for timber and firewood. By the mid-1800s farmland quilted the region, a railroad had arrived, Indian footpaths had become roads for horse-drawn wagons, and proliferating byways had further fragmented the remaining forest lots. Old Indian Path, the easternmost boundary of the original Harrison land, soon became the busy Lincoln Highway, which carried automobiles between Philadelphia and New York City.

Changes in this part of the country mirrored what was happening to forests throughout many settled regions of the Northeast. Farms were built, forests cut, and by 1800 the 4 million settlers in the Northeast had spilled into the remotest corners of New England. At farming's peak around 1900, more than half of the 100 million acres of northeastern forests had been cut, including all but a fraction of Harrison's original land. By that time the largest stand of original trees in his twenty-seven-square-mile purchase was a sixty-five-acre woodlot. That forested enclave must have been a spectacle even in the mid-1800s: two- and three-hundred-year-old trees presiding over a shadowed realm of birdsong, butterflies and flying squirrels, grouse, turkeys, and perhaps a bear or panther passing through on its way inland or farther north.

In the mid-1950s this final trace of the original forest came under assault from a timber company hoping to liquidate the valuable hardwood. To protect this rare jewel of nature, in 1955 several organizations, including the United Brotherhood of Carpenters and Joiners of America, purchased the land, named it in honor of a former union president, and donated it to nearby Rutgers University. Today, hidden between sprawling New Brunswick and Somerville, New Jersey, the sixty-five-acre William L. Hutcheson Memorial Forest remains one of the largest old-growth oak-hickory forests in the mid-Atlantic states.

2.

On a hot morning in June 2001, three hundred years after Harrison's purchase, I drove to Hutcheson Memorial Forest. There I was greeted by the forest's director, Edmund W. Stiles, a Rutgers ecology professor. After we introduced ourselves, we walked down the path into the ancient grove.

The air turned cooler as the bright morning light dissolved into the soft hues of the forest edge. There seemed to be as many fallen trees as standing ones. Although not the massive, moss-draped druids of the purple prose often used to describe a primeval forest, they were the biggest trees I had ever seen in New Jersey. Their massive branches created a heavy latticework against the blue sky.

"Ecologists once believed that forests reached a climax and would stay that way," Stiles began. "Maybe that's where the notion of the forest primeval arose. But it's not like that. About 250 years is the age of old trees in this patch. The oldest tree ever recorded here was 344 years old, from 1611. It was blown down in a hurricane in 1955."

Stiles said that since the forest's beginning—sometime after the last ice age, 10,000 years ago—natural catastrophes, especially epic storms, had struck every few centuries. Truly ancient trees aren't seen there because they get knocked down. Put another way, the forest is ancient, but the trees are not.

"One way to talk about an old-growth patch like Hutcheson isn't in terms of the age of the trees but in the length of intervals between major natural disruptions. It's not just about the trees but about the process, the whole system. Another way to think of an old-growth forest is as a place where trees die natural deaths rather than getting cut down."

Is a healthy forest as much about dead trees as live ones? I asked. Stiles nodded and explained that a dead tree provides an opportunity for numerous insects, birds, and mammals to contribute to the forest for hundreds of years. When a tree falls, mosses and other plants colonize it. Even the hole left in the ground by upturned roots becomes new habitat for small, enterprising species. But it takes a long time for trees to die and begin to return to the soil—a luxury of time that many of today's forests don't have. "It disappoints me when forest managers talk about having to clean out 'dead wood,'" Stiles said. "In cleaning out dead trees you destroy habitat that makes a healthy forest."

A severely disrupted forest, Stiles continued, can quickly lose many of its most "specialized" species—animals that can't quickly adapt to new habitats or sources of food. "Generalists," on the other hand, often accommodate change. While specialized species vacate the forest, the resourceful generalists, such as deer and mice, often expand their numbers.

"The age of trees is a huge influence on the animal community, and one of the many ways eastern forests have been degraded is by keeping them young. Another way is by fragmenting a forest into patches, like here at Hutcheson. If a forest is under a certain acreage, many animals can't live there."

There is almost as much forest in the East today as there was two centuries ago, but its pattern is quite different now, Stiles went on. After peaking about 1900, eastern farms declined as western trade routes opened markets to cheaper midwestern grain. Where the farms were abandoned, trees often grew again. By the early 1900s forests had returned to cover 50 million acres, and today forests cover three-quarters of their historical range. But one should not be fooled by size alone: even though the trees may have returned, the forests have not. Farms flowed in with people and livestock and then washed out, taking with them mountain lions, wolves, bison, wolverines, elk, mountain lions, bobcats, fishers, and numerous other species.

As we continued our stroll, I commented that the forest interior was browner than I would have imagined for an ancient forest—far from the deep, leafy tunnels, green boughs, and verdant undergrowth I expected to see. Stiles pointed out that the thick canopy of the dominant oaks, hickories, and, especially, sugar maples filters light. "Availability of sunlight influences what grows and doesn't grow on the forest

. . . .

floor," he continued, explaining why some old-growth forests can be remarkably park-like in effect.

A fragmented forest receives more sunlight than a contiguous one because every road, clearing, or cut is a virtual skylight. The light fosters more leafy growth along the forest edge, which provides browse for deer. That's one reason why deer do so well in fragmented forests. Where there's sunlight, there's browse, and browse attracts deer.

"Deer thrive and forest sickens," I commented, paraphrasing a recent headline in New Jersey's *Star-Ledger*. "The article said that white-tailed deer were almost extirpated from New Jersey a hundred years ago. But now they number in the hundreds of thousands."

Stiles snorted. "Those deer are living on borrowed time," he said. "Three hundred years ago there were probably about 25 per square mile. Now there are something like 200 per square mile. Something will knock them down, and it could be disease."

He paused on the trail and placed his palm against the bleached skeleton of a dead oak. As if to highlight the contrast with other forests in the region, Stiles explained that the dead tree had been standing when he first came to this patch, more than thirty years ago. When the next generation of biologists comes to this forest, the toppled trunk will, perhaps, have disintegrated and become little more than a raised ridge in the ground. Yet it will still be contributing to the forest's health.

Stiles pointed toward the forest edge as we continued on.

Through the brush could be seen a distant cultivated hillside and a row of houses on the horizon. I pressed the soles of my shoes hard against the never-tilled earth. The roots below clutched soil and boulders made from glaciers that had retreated 10,000 years ago and the stone arrowheads of ancient hunters who had passed through. As I gazed upward, the outstretched boughs of an oak seemed to embrace and hold me. What threads we silently break; what voices we still. By what grace, I wondered, have we been kept so well by what we have abused for so long.

3.

Like much of the Raritan River valley, Hunterdon County, New Jersey, has seen some of the most rapid development in the East. Drained by three graceful rivers, cloaked by beautiful, if young, forests, and within commuting distance of New York City, Hunterdon has been transformed over several decades from countryside to a suburbanized hub with more than 125,000 people.

"When I got here in 1985, I thought I was coming to a quiet rural county," said John Beckley, who lived at the time in the town of Annandale, about two miles from a branch of the Raritan River. As the county's director of public health, Beckley had seen firsthand many of the changes wrought by population growth and commercial development. "In one respect we're no different from many other places in America. It's just happening a lot faster here," he said. "When I got here I expected to deal with bread-and-butter public health issues. We had an environmental staff of four and spent most of our time inspecting or issuing permits for septic systems and wells—a dozen or so a week. Since then, my job's gotten a lot more interesting."

In 1985, only months after Beckley arrived at his new job, New Jersey reported its first case of infection with the human immunodeficiency virus (HIV), which had been isolated and described two years before. In 1986 the county had its first cases of another new illness, Legionnaire's disease. Two custodial workers, who survived, were stricken at Hunterdon Central Regional High School, Beckley said. Eventually, the Centers for Disease Control and Prevention discovered that bacteria living in a poorly designed water heater had caused that particular outbreak.

Next, in 1989, Beckley's office began receiving reports of raccoons behaving strangely, wandering across the county's highways and into people's yards to attack their dogs. "It turned out to be the first outbreak of terrestrial rabies in the state in nearly half a century," he said.

Ten years later West Nile virus arrived, an event that led to the county's first mosquito-control program. "We have a whole lab now, several trucks, a special freezer for preserving specimens at minus 70 degrees, and a mosquito-control team," Beckley said. "You can imagine how our staff and budget have grown since I got here in 1985. None of the traditional public health challenges have gone away. The fact is, we have more infectious diseases than before."

Many of the diseases that suddenly struck Hunterdon

County were not random: they were precipitated or fostered by human changes to the environment. The rabies outbreak, for example, was traced to hunters who transported raccoons from Florida and released them locally to improve hunting farther north, in West Virginia. Some of these raccoons were infected with the rabies virus. From there the virus marched north through the species, right into New Jersey. Legionnaire's disease, a technology-related illness, was caused when the ubiquitous *Legionella* bacteria were given the opportunity to collect in warm environments provided by modern life, such as water heaters, saunas, and air conditioners, and were then aerosolized and inhaled by people nearby.

"All these diseases were occurring against the backdrop of what has become our single biggest infectious disease problem," Beckley continued. He was referring to Lyme disease, the most common vector-borne illness in the United States. It's another disease that accumulating evidence indicates has emerged in part because of radical changes people have made to the landscape—in this case the once comparatively stable and biologically rich forests of the eastern United States, of which Hutcheson Memorial Forest is now only a sad token.

According to Sarah E. Randolph, a professor of parasite ecology at the University of Oxford, England, "it isn't known when the Lyme disease bacterium was first introduced into the United States, but it is difficult to believe that it is as recent as the last major resurgence." In other words, the bacterium that causes Lyme disease very likely has been in the United States for a long time, but until recently the conditions did not exist for the disease to become epidemic. Randolph said that though "no one can say for sure when the disease first appeared in the UK, the bacterium has been around for a long time, at least in Europe."

In the United States, Lyme disease was first described in Old Lyme, Connecticut, in the 1970s, but it wasn't documented in Hunterdon until 1988, when 12 patients were identified. There were 30 cases in 1989, and by 1993 there were 204. In 2000 Hunterdon had more than 500 cases. There are so many cases of Lyme disease that Hunterdon County continues to jockey for first place nationally with Nantucket, Massachusetts, and Columbia County, New York.

Nationally, the disease also continues to increase, although evolving case definitions make the numbers hard to accurately estimate. In 2009 the Centers for Disease Control and Prevention reported about 39,000 confirmed and probable cases—with the real number of infections as much as ten times higher.

With Hunterdon County near the top of the list, the CDC sent a team to investigate—Beckley was one of the members—and concluded that one reason for the high incidence in Hunterdon at the time was the county's high density of deer, which harbor the tick that carries the Lyme disease bacterium, near residential areas. Another was the high number of rock walls and woodpiles near homes. These provide refuge and breeding grounds for mice and chipmunks, which also carry the ticks. "Where the edge of a yard comes up against the woods, that's an 'ecotonal edge,' which is perfect habitat for ticks," Beckley explained. "That nature-culture border is where people, who may be mowing the lawn or trimming branches, often pick them up. Human activity has put people right at the center of the tick's life cycle."

Deer and other large mammals, birds, and small rodents such as mice are the literal lifeblood of the ticks. These mammals provide not only blood meals but also a means of transportation and dissemination for the otherwise largely immobile ticks. The life cycle begins in fall, when the egg-laden females drop from the animal carrying them to the ground, frequently nestling in leaf litter for the winter. With the advent of warm spring weather, the eggs hatch and the larvae hitch a ride on mice, chipmunks, or any other small mammal or bird nearby. Once on a host, the ticks feed for several days and then drop off. They develop over the next several months and re-emerge as nymphs the following spring. By then they have become mobile enough to climb low-lying bushes, where they often perch at the end of a branch or leaf and wait for another host, sometimes a larger mammal, such as a deer, to pass by. A horse, dog, or human will do. It is by these poppy-seed-size nymphs that most people become infected.

Hunterdon County apparently didn't even have deer ticks until the mid-1980s, according to Beckley. Or at least not enough to notice. An increase in deer numbers and perhaps a warming climate may have contributed to the increase.

Over the past century, the average temperature in nearby New Brunswick increased by almost two degrees and precipitation increased in that part of the state. These climatic changes may have helped to create ideal conditions for the ticks. The changes also generally paralleled an explosion in tick populations throughout the northeastern and upper north-central states.

What's more, in Hunterdon County, ticks found an estimated 30,000 deer to feed on—more than in any other county in the state. "Hunterdon County may be God's country, but it's also tick country. At least now," Beckley quipped.

There was some hope that hunting would reduce the deer population—and, thereby, the population of ticks, or so the commonly held theory went. "But most hunters want to shoot antlered bucks," Beckley explained. "Because the bucks are polygamous, even if their numbers are reduced markedly, most of the remaining females will still likely get pregnant, ensuring a high birthrate the following year." In an effort to tip the balance, officials of the New Jersey Division of Fish and Wildlife initiated an "Earn a Buck" program whereby a hunter who kills an antlerless deer—presumably a doe—can then legally shoot a buck.

Rapid development is quickly neutralizing the potential benefits of hunting in reducing deer, however: it's illegal to hunt within 450 feet of a residence without the owner's permission, and many open tracts of land are privately held and not accessible to hunters. The new houses going up in the county are therefore creating more safe havens for deer. All the environmental, economic, social, and political dynamics are thus weighted in favor of deer herds living in proximity to humans, and this increases the risk of people getting Lyme disease and other tick-borne illnesses.

Experts have considered other approaches to reducing the deer population, including netting herds and then humanely killing them, using sharpshooters, or even instituting birth control. But these are expensive or unproven solutions, and discussion of them frequently riles up animal rights proponents. "Politically, bringing about a significant reduction in the deer population is a very difficult goal to accomplish," Beckley said.

But killing deer may not be as effective as people would hope, or, for that matter, as effective as some early studies suggested. The virus-carrying ticks have alternative pathways for reaching people.

For one thing, if the number of deer were reduced, each deer would end up carrying a lot more ticks, with the total sum of ticks remaining very high. Moreover, if every deer in the forest were one day gone, the ticks would simply shift to feeding on the ample supply of other mammals, especially mice and chipmunks. Deer may come and go, but the ticks that cause Lyme disease would remain, it seems, forever.

Still, Lyme disease should be easily preventable. Prevent tick bites and you prevent the disease. For many years, the Hunterdon County Department of Health has had Lyme disease education and awareness programs in place. Several full-time staff members, including a health educator, are at work on these efforts. "Despite stressing to the public the importance of prevention—use insect repellant, avoid tick areas, stay on trails during tick season, wear light-colored clothing, and check yourself carefully when you come inside—we still can't seem to decrease the county's infection rate," Beckley said.

4.

"In early October 2001, John and I were driving home from a weekend on Cape Cod and I started feeling this really pronounced stiffness in my spine," Linda began. "All my muscles hurt. I lost my appetite and got very agitated. I got a fever, and shivers came in spasms." The night after a nurse practitioner diagnosed her illness as flu, John Beckley noticed the telltale bull's-eye rash on his wife's right shoulder blade. She had Lyme disease, he felt certain. A visit with her doctor and a three-week course of antibiotics cured her symptoms. She was lucky to have been quickly diagnosed; many people don't realize they have Lyme disease until the symptoms are far worse, and in some cases permanent, including painful joint or neurological damage.

Linda Beckley told me she wasn't sure where she picked up the tick, but she believed it happened as she was walking their dog, Willie, near the South Branch of the Raritan, not far from home. Twice before, she had found ticks in her car after such walks.

We walked outside to the back porch. As with many homes in the community, the Beckleys' backyard is carved out of the woods. Earlier, I had driven through this urban archipelago, passing islands of lawn in a sea of fragmented forest. Wooded peninsulas wrapped behind houses and extended into front yards. Paved roads ran around the neighborhood in the maze-like geometry of an integrated circuit board. Grass abutted woodlands everywhere, and the deer, squirrels, and chipmunks I saw that afternoon readily crossed between both. Ornamental shrubbery lined the foundation at the front and back of the Beckleys' house. It was dream habitat for deer, just as the wood stacks along driveways and rock walls were ideal harbors for mice and chipmunks.

Linda and I drove toward the park to walk Willie. "When I moved here three years ago, all this used to be a big farm," she said, sweeping her hand above the steering wheel as we left their neighborhood and entered a new housing development. "When John got here, beyond the farm was all woods. Now it's all these new houses."

With the houses came legions of people suddenly thrust within arm's length of the deer that came to feed on the lawns. Along new clearings, leafy browse flourished. Rock walls were built at the perimeters of properties, and woodpiles appeared at the edges of driveways, creating a paradise for the carriers of Lyme disease ticks.

When we got to the park, Linda opened her door, and before she had even unrolled the leash, Willie barreled out and romped across a field toward the river. As we walked toward him we passed a wooded area, catching a glimpse of five deer in chocolate-brown coats. Willie, meanwhile,

· · · · · · · ·

had swum across the South Branch and was heading up an embankment beneath towering cement pylons supporting Interstate 78, where four lanes of traffic roared above the opposite bank.

"Get back here, Willie!" Linda called. While we waited, I bent down, picked up a dried gray branch, and tossed it into the river. The current grabbed it and swirled it around. Willie finally returned and scrambled up the wooded stream bank toward us.

"He does this sometimes," Linda said apologetically.

On the drive back home to Madison, New Jersey, that evening, I was delayed by the aftermath of a traffic accident on I-78. As I waited at a standstill, I imagined how far along its journey toward Hutcheson Memorial Forest the stick was that I had thrown into the South Branch of the Raritan. And that brought me back to the June morning several months before when I spent time in the old-growth forest with Edmund Stiles. Age is defined not only by objects such as trees but also by the subtle processes of a growing forest, I thought. Where, I wondered, were the long, uninterrupted interludes of our world—a world in which intervals between major changes seem to shrink ever smaller day by day: neighbors moving to new jobs in other cities, a mall newly constructed here, a farm giving way to a new housing development there.

Even the intervals of Linda's disease had been shrunk, in a sense. Her doctors had compressed the definition of her Lyme disease into the interval between when she was bitten and when she successfully completed her course of antibiotics.

And she was one of the lucky ones. In as many as one in four cases the infection does not announce itself with the trademark bull's-eye rash. This means that some victims go untreated. More complicated yet, even some of those who are treated early, like Linda, may still develop debilitating symptoms years later: painful muscles and joints, depression, exhaustion, and chronic brain-fog.

Whatever the clinical course of the disease, this medical definition of Lyme disease invariably excludes the larger ecological implications of the illness and therefore its full meaning. Linda's illness was not just about a bacterium that entered her body. It was an extension of the unfortunate history of the eastern forests, and it was connected to autumn oaks and hickories, an absence of predators, and an overabundance of deer and other small mammals, such as mice and chipmunks. Her illness was not exclusively hers. It was an intimate part of a picture almost too big to see.

5.

Perhaps no one understands this big picture—the ecology of Lyme disease—better than Richard Ostfeld, an ecologist with the Cary Institute of Ecosystem Studies in Millbrook, New York. In the mid-1990s, Ostfeld began to suspect he might predict people's risk of contracting Lyme disease based upon, of all things, the abundance of acorns in a region. Acorns come in bursts, or "masts," with almost none produced in some years and bumper crops produced in others. These cycles are synchronized among trees over large regions of the country, in part by regional weather.

Ostfeld reasoned that if acorns attract deer and mice and the incidence of Lyme disease in humans is related to the densities of these animals, then the rate of human infection could be related to the production of acorns. Ecologists call it a cascade effect.

The year 1995, a very poor one for acorn production near Millbrook, might give him and his colleagues an opportunity to test his theory. Millbrook is in Dutchess County, which has an unusually high rate of Lyme disease. With lots of oaks and people, it was an ideal place to conduct the study. Ostfeld and his team measured and demarcated two sets of plots in the forest at the institute. On half of the plots, Ostfeld let the poor natural acorn crop fall. On the other plots, he supplemented nature's production with nearly a million acorns from elsewhere. In the months that followed, he regularly visited the plots and compared what happened in the supplemented plots with developments in the acorn-poor ones.

For one thing, he noted, the supplemented plots attracted far more deer that fall. Logically, this could trigger the cascade leading to a surge of Lyme disease later on. The following spring, something else besides more deer was evident: mouse populations had also exploded in these same plots because more of the well-fed adult mice survived the winter than had their poorer kin, and they had more young in the spring. Ostfeld and his colleagues dragged strips of fabric over the plots, a standard method for collecting ticks. Astonishingly, the acorn-rich plots had eight times as many newly hatched ticks, or larvae, as the regular plots. The acorns would not have attracted the ticks. Rather, the greater number of deer attracted by the abundance of acorns meant that more adult ticks could drop from the deer as the animals fed that fall. In the spring, the female ticks on the ground laid eggs, which hatched into a superabundance of larvae by early summer of 1996.

The abundance of acorns also led to an explosion of mice because they were better fed through the winter and had more young mice in the spring. A superabundance of mice would mean a lot more warm bodies for ticks to feed on. This also contributed to the explosion of ticks the following spring and two springs later, when the people-infecting nymphal ticks arose.

There was something remarkable about the ticks that hatched from those eggs: they didn't harbor the bacteria that cause Lyme disease. Even if the mother tick was infected, the bacteria were not passed through the eggs to the larvae. To become infected, a tick first had to feed on an infected animal, such as a mouse, chipmunk, or deer. And since almost all mice carry the bacteria (unlike some other forest animals), almost every tick that feeds on a mouse becomes infected.

It stood to reason that the more mice there were in an area, the more likely it was that actively feeding ticks in

· · · · · · · ·

that area would become infected. And since more mice had been drawn to the acorn-rich plots, Ostfeld was not surprised to find a higher percentage of infected ticks there. Moreover, a higher density of infected ticks in an area frequented by people means a higher infection rate among people. Acorns attract deer and mice, mice infect ticks, and infected ticks give people Lyme disease. But could people's health really be linked to something as seemingly unrelated as acorn production?

One of the first tests of Ostfeld's theory would occur after a natural mast. He could then compare infected tick populations in regions where a bumper crop occurred with infection rates in places where it had not, or he could compare infection rates in the same area over several successive years. A spike in infections at a certain interval after each large acorn crop might lend weight to his theory.

6.

The year 1997 saw one of the most prolific acorn crops in years in the mid-Atlantic states. In Hutcheson Memorial Forest, Edmund Stiles recalled the abundance of acorns that year. Moving through the forest, he said, was "like walking on marbles." Perhaps no place experienced a greater rain of the acorns than the beautiful, oak-arched campus of Drew University in Madison, New Jersey. The campus is situated in a grove of old oaks, many of them towering a hundred feet overhead. The school has been called the "university in the forest."

Twenty-year-old Jeff Dunbar was a sophomore in the fall

of 1997, and so many acorns pummeled his dorm roof that term that the noise sometimes woke him at night. By day, deer from a nearby forest wandered across campus, and chipmunks, squirrels, and mice scurried amid the landscaping at the building's foundations. The presence of so many deer and small mammals on campus that fall meant that numerous egg-laden ticks were falling to the ground. And two years later—by the summer of 1999—these led to a population explosion of blood-hungry nymphs. That explosion happened to be the very summer Dunbar decided to live on campus while working for a state assemblyman nearby. Dunbar played Frisbee on campus almost every evening, and wayward throws often sent him scrambling through tick-laden bushes.

On August 1, Dunbar awoke with the left side of his face paralyzed. He went to the emergency room at nearby Morristown Memorial Hospital, where the attending physician treated him with steroids. A Lyme test, which measures the body's immune response to the Lyme disease bacterium, was negative. But the test can be inconclusive if the body's immune response has not yet kicked in. The disease can therefore go undetected. His facial paralysis disappeared two weeks later without further treatment.

He continued his job into the winter of 2000. That February, as he was stuffing mailings into envelopes, his shoulders and elbows became so stiff and sore that he could hardly move them. His physician concluded that the repetitive stuffing motion had strained both shoulder joints, and he recommended physical therapy. Dunbar improved, but he still felt tired and sore. During the summer of 2001, as he was undergoing routine arthroscopic surgery for a slight tear in a knee ligament, his surgeon discovered severe inflammation in the joint. A Lyme test ordered by the surgeon was positive. By now, the bacteria had invaded Dunbar's joints and spinal fluid. After eight weeks of intravenous antibiotics, he improved, and his symptoms largely disappeared.

Meanwhile, Richard Ostfeld had gathered data on the rate of Lyme disease in the mid-Atlantic states, where the large acorn mast had occurred in 1997—two years earlier. If his theory was correct, the rate of infection should rise among people there in the second year after the mast. The 1999 infection rate did increase in the area affected by the mast of 1997. In fact, 1999, the year Dunbar became ill, saw the third-highest number of Lyme disease cases ever reported in the mid-Atlantic region.

Although Ostfeld was encouraged by the results, this was far from proof of his hypothesis. Over the ensuing years he collected vast amounts of data on acorn production and the abundance of deer, mice, and chipmunks. Although lots of acorns in the fall attracted lots of deer from other areas of the forest, a single acorn mast didn't necessarily lead to a population boom in deer. Their long generation time meant their populations didn't respond quickly. Mice and other small rodents, however, which reproduce more rapidly, did experience a population boom. In other words, mice, not deer, were the strongest link between lots of acorns and an increase in the number of infected ticks—and the increased risk to people—two years later.

"Although people frequently blame deer populations for increases in Lyme disease, increases in deer populations past a low threshold don't lead to an increased number of the nymphal ticks that transmit Lyme disease to humans," Ostfeld said. "But increases in mouse populations do lead to greater densities of these ticks. It's almost axiomatic that deer are always the center of the tick life cycle, and it's very hard to convince a lot of people otherwise. But that's not what our data show. It's the mice."

7.

Whatever the disagreements among experts, Ostfeld's research has shown a link between acorn abundance and the risk of people coming into contact with infected ticks.

Mice and chipmunks transmit Lyme disease to more than 90 percent of ticks that feed on them, whereas opossums, raccoons, deer, birds, and many other forest dwellers infect only about 10 percent or less of their ticks. This contrast goes to the heart of the ecology of Lyme disease. The feeding options for ticks increase in almost direct proportion to the variety of species they can choose from. In theory, a forest with a greater variety of animals—as, say, had once been present in Hutcheson Memorial Forest, which I visited with Ted Stiles—would reduce the probability of a tick feeding on a mouse. And the chance of the tick picking up the Lyme disease bacterium would also be reduced.

At the mention of Stiles, Ostfeld, who knew him, sadly told me that the renowned professor had since passed away. Stiles, he said, had done so much to preserve forests in New Jersey.

Ostfeld wondered if the loss of species in the northeastern forests, which now favored generalists such as mice and chipmunks, had in fact contributed to the increase in Lyme disease. Conversely, if a greater variety of species were returned to a forest, would that reduce the density of mice and chipmunks and therefore the high rate of infection among ticks—and people? Would a greater degree of biological diversity help to protect people from Lyme disease—that is, dilute the impact of the disease?

Ostfeld couldn't recreate the rich and diverse forests of old, but he could test the dilution theory with computer modeling. So he began by creating a computerized forest. When he added a new species to the computer-modeled forest, the density of ticks infected with the Lyme disease bacterium declined. But would this effect hold true in the field? Ostfeld decided to find out.

He and his colleagues tried to get a real-world grasp of his hypothesis by listing all the eastern bird, mammal, and lizard species, from Florida to Maine, on which ticks carrying Lyme disease are known to feed. The farther south one moves, the greater is the diversity of species. The researchers then compared the numbers of different species within regions along the eastern seaboard with the rates of Lyme disease in people in those same regions. They found that the areas with more species had fewer cases of Lyme disease per capita. High biological diversity, it seemed, did tend to minimize the rate of Lyme disease infection in the human population—at least that was one reasonable interpretation of his findings.

In theory, increased biological diversity could reduce the risk of Lyme disease in several specific ways. First, it could decrease the number of mice that transmit the bacteria to ticks. Ostfeld and his colleagues tested this by setting up forty field sites in New York, New Jersey, and Connecticut where Lyme disease was endemic. The sites represented a cross section of habitats, from low to high quality. He and his research team inventoried all the mammal and bird species in each of the sites. As predicted, the more other species there were (including predators of mice, such as foxes), the fewer the mice.

Second, increased biodiversity could reduce Lyme disease by lowering the chance that a tick would therefore encounter an infected mouse to feed on. Using vast amounts of data about the number of ticks on mice from the forest (he, his colleague Jesse Brunner, and other members of the team had captured more than 10,000 over fourteen years of study), Ostfeld statistically analyzed the factors that reduced the number of ticks on a mouse. The answer was the abundance of chipmunks in the area. The higher the density of chipmunks, the lower the chance that a tick would land on a mouse and become infected.

Mouse populations within plots that had no chipmunks had, on average, twenty more ticks than did those within

. . . .

plots that had eighty chipmunks. In other words, biodiversity reduced the chance that any given tick would encounter an infective mouse.

The third way biodiversity should reduce Lyme disease was by decreasing the number of all the best tick vectors.

Some hosts are better for ticks than others. Some animals, such as opossums, quickly groom off the marauding ticks. Only about 3.5 percent of ticks that manage to get onto possums survive, but more than half of those that get onto white-footed mice survive to eat their fill of blood. Mice are tick havens, whereas opossums are tick traps. Chipmunks, squirrels, and some bird species fall between these two extremes. Therefore, it would stand to reason that the greater the number of less-than-ideal vectors there were to attract ticks, the fewer ticks would survive.

From all this, Ostfeld finally could comfortably draw the conclusion that increasing biological diversity could reduce Lyme disease. And there was one sure way to increase this protective biological diversity: decrease forest fragmentation.

Larger forests mean more predators and, therefore, a lower density of mice. Larger forests mean more species to dilute the Lyme disease bacterium. And larger forests lower the chance that a tick will find a species that will easily transmit the bacterium to it.

"Small forest fragments of five acres or less are really risky places," Ostfeld said. "In those small fragments, risk for Lyme disease goes up fourfold, compared with that in forests of twelve to twenty acres."

"Fragmentation favors good tick hosts and reservoirs, like shrews, mice, and chipmunks," he said. "Fragmentation disfavors predators, such as foxes, that prey on the small hosts. Foxes live in family groups, with moms foraging for mice and other small rodents to support their big litters of pups for six months or more. Foxes naturally have higher densities than coyotes in healthy forests. But when coyotes move into fragmented areas, they displace and kill foxes. The incidence of Lyme disease then goes up."

8.

The ecology of Lyme disease reminds us that many connections between the health of the earth and human health are deeply woven into ecology. Changes in forests and their species, research such as Ostfeld's suggests, are reflected in human disease. Places such as William L. Hutcheson Memorial Forest are touchstones of seeming tranquility in a world undergoing constant change caused by humans. We will probably never know if Lyme disease afflicted forest dwellers there five hundred years ago, but the diverse ecology at that time would have weighed against it. If the Lyme disease bacterium were present, the indigenous forest dwellers might, over millennia, have developed immunity to it. What we can be sure of is that, in our overzealous efforts to make the world more hospitable for humans, we have been making it more hospitable for some of the microbes that make us ill.

A Spring to Die For: Hantavirus

1.

The mysterious illness that killed two young Navajos
in the spring of 1993—and the deaths that followed—
captured the nation's attention like no other outbreak
since Legionnaire's disease in 1976. The victims, Merrill
Bahe, 20, and his fiancee, Florena Woody, 21, were
young and had no medical history that might explain
why they had become ill. The doctors who treated them
had never seen anything like it.

Merrill Bahe and Florena Woody grew up in two
starkly different worlds on the same Indian reservation,
25,000 square miles of land extending into New Mexico,
Arizona, and Utah.

For Merrill Bahe, each day brought a new struggle
with poverty. Each weekday morning, Merrill would

awaken at 5 a.m., slip out of his blanket—the only bedding he had ever known—and creep toward the door of his family's wood-and-tar-paper shanty.

It was as much to ease the burden on his family as to get a square meal that Merrill Bahe would time his hour-long run so that he would arrive at Torreon Middle School two hours before his first class, in time to eat breakfast with the kindly kitchen staff. One day, Torreon's track coach, impressed with the boy's speed and strength of character, called track coach Mike Gorospe at the Santa Fe Indian School, a boarding school set up by the U.S. government and now run by the 19 Pueblo tribes.

Merrill Bahe's acceptance to boarding school for the fall term in 1988 promised to change his life.

Florena Woody had always been lively and enthusiastic, eager to ride spirited horses and climb the cliffs behind the family's trailer encampment in Littlewater, N.M., to pilfer feathers from an eagle's nest. Now she had a new radiance. Florena's brother Collins remembers that every time the talk turned to Merrill Bahe, Florena Woody "just lit up."

Florena Woody's illness began on April 29, 1993, with nothing more alarming than muscle aches in her neck and shoulders. Four days later, fever set in. She began to cough. On May 6, Florena visited her doctor, who told her it was probably a mild case of the flu. The doctor gave Florena a shot and antibiotics. Nothing seemed to help. One week later, Florena was still feverish. That

Saturday night it became clear to Bita Begay that her daughter was becoming increasingly lethargic. She decided to take her to Crownpoint Hospital, seven miles away. "I can't breathe," Florena Woody lamented, as the attending physician examined her.

The doctor ordered a chest X-ray. Soft tissue normally appears black on X-rays. [The film] gave him a "sinking feeling." Florena Woody's lungs were white. Her lungs were rapidly, and inexplicably, filling with fluid.

From their vantage point in the hallway, the Woody family watched in disbelief as the hospital staff replaced the second bed in the room with a mechanical ventilator. But the task was futile. An alarm shrilled. Faces turned toward the heart monitor. A glowing green wave form told them that Florena's heart had stalled, then stopped.

Merrill Bahe's symptoms, mild at first, worsened two days later. It was Tuesday, May 11. Florena's funeral was just three days away.

"Go to the hospital," Bita Begay ordered. Merrill climbed into the pickup, and Florena's cousin Karoline drove him to the Crownpoint Hospital, where a puzzled young doctor found abnormalities in Merrill's blood test but nothing that revealed what was wrong. The doctors decided to discharge Merrill with stern instructions to return if his symptoms worsened.

By Friday, the morning of the funeral, Merrill's lips were turning faintly blue from lack of oxygen. Collins asked his cousin Karoline to drive Merrill to the hospital in Gallup.

About 10 miles north of Thoreau, Merrill began visibly struggling for every breath. His skin was sallow, and his lips turned a deeper blue. Karoline pulled into the parking lot of B.J.'s Convenience Store, in the tiny roadside community of Thoreau. Paramedics from the Thoreau Volunteer Ambulance Co. arrived moments later, but they could not revive him.

Patricia McFeeley, deputy director of the OMI, conducted a limited autopsy on Merrill, removing just enough tissue for the state laboratory to test for pneumonic plague, a flea-borne disease that occurs regularly in the Four Corners. But the plague tests, completed after midnight, were negative. Something else had killed the young couple. And McFeeley had no idea what it could be.

—Steve Sternberg, "An Outbreak of Pain"

2.

The Colorado Plateau, which stretches across 130,000 square miles of southeastern Utah, northern Arizona, northwestern New Mexico, and western Colorado, has seen more rain and snow during the past twenty-five years than at any other time during the past two hundred. And those two centuries have been the wettest on the Plateau in the past 2,129 years. The Plateau is actually a huge basin filled with tablelands and surrounded by mountains. It is a world apart from the rest of the Southwest, older, with its own assemblage of plants and animals and climatic patterns.

The Plateau's remarkable climatic history is told by ancient trees or their enduring remains—some more than a thousand years old. Each spring, beginning at a time lost to all but the memory of these relics, a new growth layer swelled beneath their bark. Toward the end of the growing season, as sap drained away, the layer remained, and the following spring a new one grew. Year by year, century by century, the process continued, creating a trunk of concentric growth rings. Rainy growing seasons tended to produce wide rings, and droughts created narrower ones. By extracting a straw-size core of wood and examining it under a microscope, tree-ring specialists such as Henri Grissino-Mayer at the University of Tennessee, Knoxville, can not only estimate a tree's age by counting the rings but also get a sense of the climatic patterns of its time by analyzing the rings' character and width.

There has been a dramatic long-term climatic shift on the Colorado Plateau, from desert-like conditions to, in more recent times, almost seasonal monsoons. Many climatologists attribute part of that shift to more frequent rises in ocean temperature near the western coast of South America, a phenomenon known as El Niño.

During El Niño years the surface waters off the west coast of South America become unusually warm, whereas during so-called La Niña years surface temperatures cool. These fluctuating ocean-surface temperatures can affect many aspects of the weather by influencing the amount of water that evaporates into the atmosphere and the course of high-altitude winds. The shift between El Niño and La Niña has historically been transient and mild. But in recent decades El Niño has been unusually persistent, leading to greater extremes in the weather, including increased precipitation in many places.

Because El Niño is accompanied by a slackening of the Pacific trade winds or even a reversal in their east-to-west direction, storms that normally pass over the Northwest can shift southward, dumping unusually heavy rains or snows on southern California and the Southwest.

If El Niño itself is natural, the extremes and duration of the heated Pacific appear to be something new—made worse by a warmer global climate, some scientists argue. And that, much evidence suggests, is the result of the present scale and character of human activity—the extent of automobile and truck exhaust, coal-powered generating plants, and other sources that emit heat-trapping gases into the atmosphere. According to Grissino-Mayer, who has studied the historical rainfall patterns on the Plateau, "global warming is intensifying many of the natural cycles such as El Niño. There's no doubt about it, in my opinion. The two-hundred-year period of increased rainfall also coincides with the increasing use of fossil fuels and the emission of greenhouse gases into the atmosphere."

A powerful El Niño cycle began in 1991, and early in the following year a severe El Niño–driven flood moved across the Los Angeles area, stranding fifty motorists in quickly rising waters and sweeping a fifteen-year-old to his death. The following month, winds laden with Pacific moisture pummeled Las Vegas with two and a half inches of rain, flooding streets and turning Duck Creek into a torrent. In December 1992, unusually heavy snows fell at Gallup, New Mexico, and over much of the Colorado Plateau, transforming the watercolor landscape into a monochrome photograph. Following heavy rains in the area, the Federal Emergency Management Agency declared the normally arid state a flood disaster area. During the first three months of 1993, a series of mild snowstorms interspersed with rain fell across the Plateau, and New Mexico was again declared a flood disaster area, along with neighboring Arizona.

Autumn and winter precipitation increased the soil moisture that fed the juniper and piñon woodlands of the Plateau, helping the piñons produce a huge crop of nuts in the fall of 1993. These were consumed by people, birds, and numerous rodents. Awakened by the fall rains, millions of downy chess grass seeds, scattered the preceding autumn, began to germinate, spreading their roots beneath the moist soil. When the first rains of spring arrived, the extensive roots quickly soaked up the water and gave birth to bright green seedlings with hairy leaves.

Snakeweed also burst forth, creating refuge and food for grasshoppers. This unusual bounty of energy-rich wildflowers, nuts, juniper berries, and cones led to an increased number of mice, whose reproductive cycles were triggered by their consumption of the abundant green vegetation. The heavy rains had changed the cycle of life on the Plateau.

3.

Hantavirus pulmonary syndrome (HPS), a usually fatal infection that causes victims to drown in their own fluids, was not exactly new—or at least not to the Navajo. It was caused, the elders said, by Na'ats'oosi, the mouse, and by ch'osh doo yit'iinii, a tiny invisible presence in the mouse's urine. It entered the nose and mouth and took the victim's breath away. An abundance of mice, Navajo elders said, brings the disease to the Plateau and kills healthy young Navajos. It had done so twice before, they said: once in the spring of 1919, a year after a devastating influenza epidemic struck the Navajo reservation, and again in 1933 and 1934, following unusually heavy winter and spring rains. The elders suspected that after the winter and spring rains of 1992 and 1993 the disease had returned. Florena Woody and Merrill Bahe were two of its victims.

Although the Navajo explanation for the disease apparently goes back for generations, the disease eluded detection by public health authorities until the severe 1993 outbreak was triggered, at least in part, by unusually heavy El Niño rains. Here was an illness rising and falling with rainfall patterns that humans themselves seemed to be influencing, and a haunting example of how the fates of a young couple in Stillwater, New Mexico, were influenced—if not sealed—by fluctuating ocean temperatures off the coast of Peru, thousands of miles away.

As the mysteries of the disease began to be unraveled, understanding of it grew as old Navajo wisdom blended with scientific analysis in a most unusual way. What emerged, at least to those who could hold both perspectives, was a powerfully new, encompassing view of humans not as a stand-alone species but as just one species among many in a web of climate, ecology, and intertwined fates. It was a view of a human illness whose significance transcended emergency rooms and the search for a cure.

4.

One modern interpreter of Navajo medical beliefs is Ben Muneta, a physician with the Indian Health Service in Albuquerque, New Mexico. Born on the Navajo reservation and trained at Stanford University School of Medicine, Muneta was working with the Indian Health Service in the spring of 1993, when a number of people in the Four Corners region were stricken with hantavirus.

"Most everyone seemed to be totally baffled by what this killer disease was, including the Centers for Disease Control and Prevention, which sent a team to investigate," Muneta told me in 2002. In June, Peterson Zah, the president of the Navajo, convened a meeting in Window Rock, Arizona, of Navajo healers to seek their guidance. Each healer spoke about how humans are not the dominant force in nature but instead are dependent upon other forms of life for existence. The outbreak had resulted from disharmony in the environment, they claimed, and now ceremonies were needed to reestablish harmony between patients and the universe.

"People at the meeting, including a few from the CDC, began to realize that hantavirus was not a new disease," Muneta said. "A few elders also spoke of Na'ats'oosi not simply as a mouse but as a 'thing that sucks on things and leaves a trail of saliva as it flees.' Perhaps they realized the virus might be spread by mouse saliva as well as urine. Elders have long had taboos to prevent human contact with mice."

Muneta believes that the Navajo not only had understood the basic ecology of the disease for centuries but also had designed a healing ceremony specifically for it. His belief is based on a Navajo sandpainting, which he photographed, that depicts a mouse and several medicinal plants, two of which went by the Navajo names Tl'oh azihii libáhígíí and awe'e'tsa'a'l.

Tl'oh azihii libáhígíí is a member of a group of plants known as ephedra, Muneta explained. Plants from this group contain ephedrine, a cardiac stimulant, which is also used in several over-the-counter asthma and allergy medications to open the airways. Interestingly, drugs with similar clinical properties are used today for supportive care of hospital patients infected with hantavirus.

A second plant depicted in the sandpainting, according to Muneta, is awe'e'tsa'a'l. The historical literature suggests this evergreen plant was used by some Native Americans in a cold medication and, when mixed with green branches, sagebrush, and juniper, could loosen the patient's mucus.

Among the non-Navajo people attending the meeting of healers in 1993 was Ron Voorhees, a physician and deputy state epidemiologist with the New Mexico Department of Health. "It became clear from that gathering that the Navajo knew just about everything about the virus and all we really added was a name," Voorhees told me. "Before the CDC identified the virus, the elders were saying, 'We've had this before! After wet winters. When there were a lot of mice.' That's essentially what all the research on the virus would later show. We've got DNA sequencing, so we can do all sorts of things and trace the evolutionary history of the virus. But the Navajo had the long history of observational epidemiology, which is pretty much what we used until computers made more complicated statistical analysis possible. The Navajo looked at people who got the disease and compared them with people who didn't; then they drew conclusions about how the people got it.

"Epidemiology is little more than structured observation," Voorhees continued. "They did their own risk-factor analysis. Navajos have a highly evolved culture, in which careful observations over many generations add up to a substantial knowledge base. One reason, probably, why they reached the same conclusion with fewer tools is that they have a much broader view of interconnectedness than we do. They are far less dependent on rigid linear connections, and they see connections in daily life that we can see only through statistics. They especially understood the basic ecology of the disease, something the CDC and the rest of us had no notion of until we heard the elders speak at the meeting."

. . . .

5.

Robert Parmenter, a professor of ecology at the University of New Mexico and director of the university's long-term ecology research program at Sevilleta Research Field Station, doesn't put much stock in Navajo claims to know so much about the disease, but he puts a lot of stock in modern science and its conclusions about the origins of hantavirus. As leader of a decade-long study of deer mouse populations in the Four Corners area, he has a lot of science to take stock in.

"The mouse data we had been collecting at our facility south of Albuquerque turned out to be invaluable because it showed fluctuating mouse populations over a very long time," Parmenter explained during my visit to the campus. "The spring of 1993 saw a huge explosion in populations. In an average year, perhaps one or two of every ten box traps we set out would catch one of the rodents. But in the spring of 1993, 90 percent of the traps were full by morning. We also kept very precise weather data, so it was easy to demonstrate, vis-à-vis the occurrence of hantavirus, that mouse populations always increased after unusually high winter and spring precipitation."

A second site, on the Navajo reservation nearly two hundred miles north of Sevilleta, had seen unusually heavy rains, and surveys there showed an increase in mice as well as in cases of HPS. A third site, in Moab, Utah, turned out to offer a scientifically convincing point of comparison because it had not rained there during the year before the outbreaks elsewhere. In Moab, the density of mice remained comparatively low, and no cases of HPS had been reported.

"You can draw three conclusions from this data," Parmenter said. First, the rains in 1992 and early 1993 caused a dramatic increase in mice. During the appearance of El Niño in the summer of 1991, deer mouse population densities in New Mexico increased from about 15 mice for every ten acres to more than 75 per ten acres eight months later. By the spring of 1993, there were about 100 mice per every ten acres. Second, the initial human cases of HPS directly followed these increased densities of mice. "If you put these two together," Parmenter went on, "you come to the conclusion that increased winter and summer rain is associated with outbreaks of hantavirus."

"What caused all the unusually heavy rains?" I asked.

"El Niño. When the rains came, so did the sickness. When the rains left, the sickness left too."

But the picture turned out to be more complicated. In 2000, researchers from Johns Hopkins University completed a more precise analysis of precipitation data during the El Niño years. They discovered that even though rainfall was above normal in many areas, it was normal around the homes where the victims became infected. The rain nevertheless played a critical role, Parmenter explained. The deer mouse populations had exploded in the areas with unusually heavy rain, and then the mice spilled out of the canyons and traveled into secondary habitats, such as around houses, trailers, outhouses, and other places—the very places where people became infected.

By the autumn of 1993, the snakeweed had become mounds of saffron flowers across the Plateau. The HPS outbreak there seemed to have vanished as quickly as it had arisen. The slender stalks of the downy grasses bent in the autumn breeze as their seed heads faded from green to purple and then brown and the plants approached their winter death, illustrating, in a sad and incongruous way, what the Navajo have always said: "In beauty it is done; in harmony it is written. In beauty and harmony it shall so be finished."

6.

As researchers studied the disease, they discovered what the Navajo already knew. As their traditions told, the disease was old, perhaps even ancient, at least in their culture. A later analysis of preserved lung tissues showed that a thirty-nine-year-old Utah man had probably died from the disease in 1959. Still more research proved that in 1978 a man, also from Utah, had died from the disease. But the scientifically documented history of hantavirus stretched back earlier—at least outside of the United States.

The name comes from the Hantan River, which flows through areas of Korea where the virus is endemic and where American soldiers and scientists were first exposed to it during the Korean War. During that time, thousands of United Nations soldiers came down with something called "Korean hemorrhagic fever." Hantavirus was later identified as the cause.

Soon after the Four Corners outbreak, CDC scientists

identified the virus as related to strains found in Europe and Asia. But once out from under the microscope, the US strain was utterly different. For one thing, it was the first known occurrence outside of Eurasia. Second, the Eurasian strains didn't cause respiratory failure. Finally, the Four Corners virus was five times more lethal than those from Europe.

Although the disease itself was not new, the way it suddenly emerged as an epidemic in the United States for the first time surely was.

Continuing research has strengthened the connection between increased populations of mice and human hantavirus infections. According to the CDC, there were ten times as many of the mice in 1993 than there had been the year before—thanks to the drought followed by heavy snows and rain, which led to massive plant growth and subsequent food for the mice, whose numbers exploded. Major outbreaks that followed also seemed to follow times of unusually wet weather.

By the summer of 2002, a total of 318 cases of the newly recognized hantavirus pulmonary syndrome had been identified in thirty-one states, including several in the Four Corners area. More than a third of the victims died. The case count peaked again in 2000 and 2006, with forty-three and forty-one cases, respectively. Again, the disease claimed the lives of more than a third of those infected.

Then, in 2012, would-be travelers to the popular Yosemite National Park in California got a jolt when the World Health

Organization warned that hantavirus pulmonary syndrome had stricken some visitors to the park.

The first cases were two Californians who had camped that summer in the Curry Village area. One died. By the end of August, four more cases had been identified. By the time the outbreak ended, ten people had become infected. Three of the cases were fatal. Nine of the victims had stayed at the Signature Tent Cabins in Curry Village, while the other had probably picked up the virus while hiking in the High Sierra Camps, about fifteen miles from the Village.

In various other regions of the United States, twenty additional people became infected, bringing the total count to thirty that year. Forty percent of the victims died.

The disease was yet another instance of what the Navajo had long known: human health and the fate of the environment are inseparable. With changes in climate and ecology, another ecodemic had arrived.

A Virus from the Nile

1.

It probably happened in August. Beyond that, no one can say when the tiny brown wisp settled upon Enrico Gabrielli's body. The sixty-year-old cherished summer evenings among the red geraniums and purple cosmos in his garden, in the Italian neighborhood of Whitestone in Queens, New York—and never more so than in the summer of 1999. In July the temperature broke ninety-five degrees for eleven straight days—the hottest month ever recorded in the city.

On Wednesday, August 11, the gray-haired Gabrielli returned from his job at a mannequin factory in Elizabeth, New Jersey, and complained of fever and chills. His wife, Caterina, suspecting the flu, handed him two aspirin tablets and sent him to bed. He shivered and sweated throughout the night.

. . . .

By the time Gabrielli was admitted to the intensive care unit of Flushing Hospital Medical Center the next day, he was feverish, disoriented, and unable to move. His strength rapidly faded. He began having trouble breathing and was put on a ventilator. A few days later, as he lay beneath dozens of get-well cards taped above the bed, Gabrielli opened his eyes and spoke. His 104-degree fever had broken. He had lost more than twenty pounds, which made his once-full face gaunt. Over the next few weeks he grew stronger, though he still could not walk on his own and relied on a catheter to urinate. Although he now walked with a cane, the life-threatening phase of his mysterious illness had passed, and Enrico Gabrielli, the first known victim of West Nile virus in the Western Hemisphere, had lived to tell about it.

On August 15, four days after Gabrielli began experiencing symptoms, an eighty-year-old man who lived a few blocks from the Gabriellis fell ill. Most evenings that summer, he and his eighty-two-year-old wife had sat outside their home, talking to each other and to passing neighbors. The annoying whine of jets passing overhead from nearby John F. Kennedy International Airport and LaGuardia Airport was a part of life. So was the sweet maritime scent of brine from the nearby marshes. Gray herons, egrets, gulls, and other shorebirds frequently passed over the neighborhood, making the skies above Queens a living diorama on the history of flight, ranging from the ultra-sophisticated structure of herons to crudely shaped modern aircraft. Aside from a manageable heart condition, the former World War II sergeant had been active and healthy. His wife believes that the mosquito bit him one August evening as he relaxed outside in his armchair.

Many houses in the neighborhood lacked air-conditioning. Evenings outside, always a favorite summer pastime in northern Queens, were a necessity that year. April, May, and June had been the driest stretch in more than a hundred years. Newspapers carried headlines about the drought and the heat, which killed more than a hundred people from the Midwest to the East Coast.

"Is something new and different going on with the weather?" asked science writer William K. Stevens in the *New York Times*. The article said that the heat wave was part of a fifty-year trend toward hotter summers in the region and that heat waves and droughts could become more frequent. Meteorologists attributed the drought to the naturally shifting warm-cold cycle of surface temperatures in the Pacific Ocean linked to El Niño and La Niña years, while climate scientists suspected that the increasing severity of shifts in the recent past has been caused by global warming.

The drought made misery for humans, but it benefited one of the most common biting insects in Queens. The northern house mosquito, *Culex pipiens*, often thrives during droughts. After getting a blood meal, a female mosquito deposits eggs in wastewater, which is laced with organic nutrients. Because of the drought, the city sewers had not been flushed by rain in months, creating the organically rich standing water the egg-laden females preferred. When the mosquito eggs hatched, the dry heat aboveground tended to confine the emerging insects to the humid sewers. On August 5, the first rain in weeks fell on Queens, helping to liberate the blood-seeking wisps from their subterranean lairs. At dusk they fanned across the borough.

Although they naturally prefer birds, the mosquitoes bite humans and other mammals as well. There are actually two beneficiaries of their blood meals: the mosquitoes themselves and any viruses they might harbor—West Nile virus in this case. The virus needs a living being—a host—within which to replicate. Each time the mosquito bites a bird, the virus within the mosquito has an opportunity to move into a new host. Without such living incubation chambers, or reservoirs—and a means of traveling to new ones—a virus would quickly die out because its life within any particular animal may be brief. By using a mosquito as a vector, the virus can quickly spread and become established in millions of birds through a process appropriately known as amplification. It is a diabolically effective system. Mosquitoes also pass the virus to people, where it can replicate in the brain. That was the case with Gabrielli and his elderly neighbor.

On Saturday, August 12, a week after the half-inch downpour, the elderly neighbor came in from mowing the front lawn and complained of extreme fatigue. "It was the first time he'd ever complained," his wife said. He wouldn't eat. He vomited and went to bed. The next morning, his wife expected him to be up as usual by four or five o'clock, banging around the kitchen and making coffee. Instead, he could

. . . .

barely open his eyes. When he slid his arm over her waist, she noticed that his hand was hot. "Somehow he couldn't move right," she remarked. She canceled plans to visit their daughter, who instead came to Queens that afternoon. When she arrived, she saw that her usually neatly dressed father had tucked in only half his shirt. He was laboring to speak in single syllables, and later in the day he collapsed in a chair. An ambulance drove him to Flushing Hospital, where doctors were able to revive him. He was admitted to the intensive care unit, only a few beds from where Enrico Gabrielli lay, but his liver and kidneys began to fail and he suffered a heart attack. Soon thereafter, he died. The former soldier was buried on Long Island—the second known victim and the first fatality of the mysterious disease.

By August 23, three more patients with neurological symptoms had been admitted to Flushing Hospital. Deborah S. Asnis, a staff physician and infectious disease specialist, telephoned the New York City Department of Health and Mental Hygiene to report the unusual cluster of illnesses. After discovering that nearby hospitals had admitted another five patients with similar symptoms, the health department contacted the Centers for Disease Control and Prevention. The next day, an official from the CDC's Epidemic Intelligence Service flew to Queens to interview surviving patients, comb medical records, and visit the homes of the afflicted in an attempt to identify the disease and determine how it spread.

By early September the CDC had come up with the supposed answer. The mayor of New York City held a news conference in Queens to announce that a disease known as St. Louis encephalitis, caused by a mosquito-borne virus, was responsible for the human deaths. St. Louis encephalitis, named for the city where, in 1933, it was first identified, had never before been seen in New York City. This was a public health emergency, and Mayor Rudolph Giuliani promised to "do everything we can to wipe out the mosquito population."

New York City's Office of Emergency Management set up a command post at 138th Street and 11th Avenue, near the Gabriellis' house. The city mobilized eleven spray trucks, five helicopters, and an airplane to douse the city with pesticides. Police officers cruised neighborhoods, warning residents over loudspeakers to remain inside with their windows closed. An advertising campaign called "Mosquito-Proof New York City" was launched.

New York City had last seen a mosquito-borne disease during a yellow fever outbreak in the early 1800s. Most modern New Yorkers could not grasp the idea of a common mosquito injecting people with a potentially fatal virus. A resident of a neighborhood adjacent to Whitestone also worried about the mental health of her children, who had developed a paralyzing phobia of flying insects. "If they see a fly," she said, "they think that they are going to die."

2.

Months before the first human death, hundreds of crows had begun dying in Queens. Some had been found within

blocks of Flushing Hospital. One woman found a disoriented crow hobbling in her garden. At Bayside Animal
Clinic, veterinarian John Charos treated more than fifty ill
crows. Half of them ultimately died. A security guard at Fort
Totten, a 163-acre government property in Queens, found
dead crows all over the base and likened it to "a plague."
Crows and other birds were also dying in the Bronx, across
the bay from Queens. Near 198th Street and Briggs Avenue, a passerby happened upon four dead pigeons. Forty
dead crows were found near the Bronx Zoo, where a captive
cormorant, three Chilean flamingos, a pheasant, and a bald
eagle also died.

Many people blamed the rash of bird deaths on the
drought. Ward Stone, pathologist at the New York State Department of Health in Albany, said it was the worst die-off
of crows in thirty years. According to the drought theory, the
heat had driven earthworms, insects, and other sources of
food deeper into the ground. As the crows dug, they encoun-
tered persistent toxins, such as DDT, that had contaminated
the soil a half-century earlier, when the pesticide was commonly used.

Tracey McNamara, a veterinarian and head pathologist at
the Bronx Zoo, questioned the drought theory. Crows were
hardy, adaptable, and resourceful birds, she thought; why
would a drought affect them more than other birds? Fur-
thermore, a drought would not have directly affected captive
zoo birds, which had all the food and water they needed.
The CDC's diagnosis of St. Louis encephalitis didn't explain

the birds' deaths, either, she realized, because birds generally aren't susceptible to St. Louis encephalitis.

McNamara's hunch was that a different—perhaps new—virus was responsible, even though viruses that killed both birds and people were few and far between in North America. One candidate, at least in theory, would be eastern equine encephalitis, or Triple E. This disease, which also attacks the brain, can kill not only birds and people but also horses and individuals of other species. The Triple E virus was known to be especially lethal to emus, ostrich-like birds from Australia. Yet the Bronx Zoo had a number of emus, and they remained healthy even as the other birds died during the mysterious outbreak. This fact alone all but ruled out Triple E as the culprit.

Another observation that weighed in favor of a new virus, McNamara believed, was that all the birds stricken by the disease at the Bronx Zoo were native to the Western Hemisphere. Did this mean that the virus had moved here from another part of the world and was killing only birds whose immune systems were unprepared for this exotic invader?

On September 9, two more flamingos died at the zoo. While taking blood from one of them, McNamara's colleague accidentally stuck herself with a contaminated needle. If the bird and human deaths were related, McNamara realized, the technician's life could be in danger. That day she called the CDC to ask about her colleague's exposure and to suggest that the human deaths and the bird die-offs were related. In doing so, she was discounting St. Louis encephalitis as the cause and thus challenging the CDC's diagnosis. The latter is not something a veterinarian—or anyone else, for that matter—usually does.

3.

Every autumn, clouds of white storks move over Israel as they migrate from breeding grounds in Europe to wintering grounds in Africa. A more direct flight to Africa would carry them across the Mediterranean Sea, but these heavy birds rely on thermals—currents of air that rise from warming land—to keep them aloft and carry them on their journey, forcing them to avoid large bodies of water, even at the expense of a longer migration route. The summer of 1998—a year before the virus struck in the United States—was the hottest in thirty-five years in Israel. Temperatures along the migration route regularly reached 100 degrees Fahrenheit and occasionally soared to 116 degrees. Winds gusted to thirty miles per hour. Unable to navigate the winds or endure the extreme heat, tens of thousands of the stressed birds that had hatched in Europe landed in Israel.

One flock of 1,200 birds set down at Eilat, in Israel's southern tip, near the Red Sea. Farmers in the region soon began finding dead storks in their fields. Not long afterward, hundreds of domestic geese in villages around the country mysteriously succumbed to an unknown disease. Many of them had neurological abnormalities: they could not stand or keep their balance. Israel's government tested a number of wild storks and the geese, and the brain of a dead goose yielded West Nile virus. Common throughout Africa

. . . .

and, more recently, in Europe, the virus had visited Israel in the 1950s and the late 1970s—but not again until the summer of 1998. The young European storks may have re-introduced the virus into Israel, where it then infected do-mesticated geese. Perhaps under normal conditions, even infected storks would have remained healthy—some birds carry the virus without ill effect—but under the stress of a difficult migration, the storks fell ill. But not, apparently, before spreading the virus to mosquitoes in the area where they landed. The mosquitoes then could have easily infected the goose farms. This drama in Israel was unfolding seven thousand miles away from Queens, and a year before either a person or a bird there would fall ill from the disease.

4.

On September 9, 1999, when McNamara telephoned the CDC's Division of Vector-Borne Infectious Diseases in Fort Collins, Colorado, to express concern about her colleague's accident, her call was transferred to the chief of the Epi-demiology and Ecology Section. McNamara asked whether the CDC would test blood samples she had taken from her colleague and the dead birds to see whether the same virus could be isolated from both. It was inappropriate, the official explained, for an institution concerned with human health to test birds' blood; indeed, since the viruses in birds and in people were different, the CDC thought it superfluous even to test the human blood sample. At a loss for what else to do, McNamara sent both samples to the National Veterinary

Services Laboratories (NVSL) in Ames, Iowa. Several days later, an official at the NVSL called McNamara to say that an unusual virus had been isolated from both samples. The lab could not determine exactly what the virus was, beyond the fact that it was a member of the dangerous *Flavivirus* genus. Yet flaviviruses had never been associated with bird fatalities in the United States. Had a new one arrived or an old one mutated?

Definitive identification of the dangerous virus would re-quire a secure laboratory to prevent human infection. Only a handful of such facilities existed in the United States. One of these was housed at the US Army Medical Research In-stitute of Infectious Diseases (USAMRIID), at Fort Detrick, Maryland, the military's main biological warfare laboratory. McNamara happened to have a friend who worked there, and that person agreed to test the samples. A researcher at the New York State Department of Health's laboratory in Albany also agreed to run further tests on the samples for-warded by the NVSL.

Meanwhile, McNamara telephoned John T. Roehrig, chief of the CDC's Arbovirus Diseases Branch, and told him that the NVSL had isolated something that looked very much like a flavivirus. She also pressed her concern that the bird and human deaths were linked—implying that the virus was indeed something new in this part of the world. Confirmation that a flavivirus was killing birds in the United States would be a historic and ominous find-ing. The CDC's response was still the same: the agency insisted that the human and avian deaths were not related and thus there was no logical reason for them to test bird samples. McNamara would have to await the findings from her friend at USAMRIID.

On September 23, the telephone rang in McNamara's Bronx Zoo laboratory. Several senior scientists from the CDC were on the line. Roehrig asked McNamara to ship frozen samples directly to the CDC that night. He said there had been some confusion with the samples the NVSL had sent them earlier. The callers said little else. To McNamara, this spoke volumes.

"Is it okay to be working with the virus here?" she asked in alarm.

Because she wore a mask and gloves and worked under a special ventilated hood, it was probably okay, she was told.

When the conference call ended, McNamara quickly telephoned several friends in the know and learned that the Fort Detrick lab had definitively ruled out three possible candidates for the outbreak: eastern equine encephalitis, western equine encephalitis, and Venezuelan equine encephalitis. The tests further suggested that St. Louis encephalitis, the CDC's original diagnosis, wasn't the cause either. USAMRIID had reported its findings immediately to the CDC and had continued searching for the identity of the mystery virus. Meanwhile, officials at the CDC, finally beginning to grow alarmed, agreed to test the samples McNamara shipped them.

On September 30, 1999, the CDC issued a press release

announcing it had now "made the link between the West Nile–like virus found in birds in New York City and the ongoing human encephalitis outbreak in the area," thus confirming McNamara's hypothesis. Geneticists immediately began comparing the New York strain with strains from Africa, Europe, and elsewhere to determine where the New York strain had originated. They soon found that it matched a sample isolated from the brain of the dead goose in Israel, a strain common throughout the Middle East.

West Nile virus was first discovered in the 1930s, when it was isolated from a woman living on the west side of the Nile River in Uganda. Since that time, birds migrating from Africa had spread the virus along their migration routes throughout much of the Middle East and Europe. But there were no bird migration routes from those countries to the East Coast of the United States. How, then, had the virus traveled thousands of miles from the Middle East to the borough of Queens?

5.

The thousands of Queens residents sitting on their porch stoops and patios the summer of 1999 had no particular reason to realize they lived near one of the greatest crossroads—for both people and birds—the world had ever known. Each month, some 11,000 overseas flights to Kennedy International Airport bring more than 2 million people through Queens. More than 20 million overseas passengers disembark there annually. That does not include the almost

4,000 horses and thousands of exotic birds, turtles, and fish and other animals that legally pass through JFK every year. Hundreds more animals—perhaps thousands—evade the quarantines and inspections set up to keep out imported diseases. And no one attempts to account for the numerous small six-legged, winged, or tiny crawling stowaways from the aircraft's cabins and pressurized holds and from the bodies of the passengers. If Queens is a cultural melting pot, it is also one gigantic petri dish. In its own way, Queens rivals some of the world's other great interspecies crossroads, such as Guangdong Province in southern China, the epicenter of the SARS outbreak.

Kennedy International Airport also lies along the Atlantic Flyway, a major migration route for birds flying between the Americas. In fact, runway 22L juts into the 10,000-acre Jamaica Bay Wildlife Refuge, which is visited every year by millions of birds from Mexico, South America, the Caribbean region, and the far north. Many people forget about the vast network of rivers, wetlands, and shorelines that surrounds New York City. The birds have not.

In spring and early summer, the songs of warblers, vireos, swamp sparrows, goldfinches, and eastern bluebirds arise from the trees and forest patches, and snowy egrets, black-crowned night herons, sandpipers, belted kingfishers, and great blue herons fill the wetlands. The occasional great cormorant, green-winged teal, and red-breasted merganser wander through. In the winter of 1998, a rare sighting of a European widgeon, perhaps from a far-flung Iceland flock, was made in Queens. Escaped parrots and other tropical birds are also occasionally documented in the borough.

Birds and people are not the only species passing through the region. Monarch butterflies migrate through New York City in autumn, feeding on life-giving milkweed in wayward urban lots and along roadsides. The monarchs are destined for New Jersey's Cape May, where they congregate by the thousands before continuing their patient journeys, on breezes or one wing-stroke at a time, to Mexico.

6.

In late August 1999, when Tropical Storm Floyd dumped nearly five inches of rain on New York City, the historic drought of that year became a memory. By mid-September, when the year's final case of West Nile fever was diagnosed, seven of the fifty-nine people hospitalized with the virus in New York City had died. Enrico Gabrielli, home from rehabilitation for several weeks, walked with a cane.

But a large part of the West Nile virus mystery remained. How had it arrived in the New York region in the first place? Perhaps a person bitten by an infected mosquito in the Middle East had carried the virus to Queens, only to be bitten by another mosquito in New York. Perhaps that mosquito then fled across the parking lot outside the international arrivals terminal and disappeared into the refuge at Jamaica Bay, where it spread the virus to other birds and mosquitoes, many of which could have ended up in the neighborhoods of Whitestone and Flushing. Perhaps an infected mosquito arrived in an aircraft cabin or cargo hold. Or maybe one of the numerous parrots, parakeets, or lovebirds smuggled through New York each year was infected. It is conceivable that, by some strange anomaly of normal migration, a bird infected with the virus in Europe or in Africa passed the infection to a bird that migrated to Queens. Given the right conditions, a single infected bird could pass the virus to a mosquito. The insect, in turn, could quickly infect other birds, igniting a rapid outbreak of the virus in both birds and people.

As autumn arrived, many birds left the New York region for their wintering grounds. A world away, a new skyful of white storks rode thermals above the hot sands of the Middle East and Israel toward Africa, as they have done since before the time of Abraham. The last wave of monarchs, propelled by the laggard storm winds, departed New York City and environs for forests in Mexico three thousand miles away.

Migrating monarchs, white storks en route to Africa, cooling waters off Peru, winds across Arabia, an empty lawn chair in Queens, and a fresh grave on Long Island. A black-crowned night heron lifted from the waters off Whitestone, circled as if on a designated flight path, and disappeared into the night.

7.

Many of the birds that left the New York City region in the autumn of 1999 migrated south along the Atlantic Flyway, some fanning into wetlands in South Carolina, Georgia, and

Florida. Although the virus was especially lethal to crows and jays, more than a hundred other species carry the virus. Many of these may have survived an infection, enabling them to carry what remained of it far and wide.

In July 2001, the first confirmed cases of West Nile fever occurred outside the New York–New Jersey metropolitan region when seventy-three-year-old Seymore Carruthers of Madison County, Florida, fell ill. A short time later, a sixty-four-year-old woman, also from Madison County, came down with the disease. "The virus is spreading," commented Steve Wiersma, chief of Florida's Bureau of Epidemiology. "We might slow it but we can't stop it. Nothing can stop it. The ecology is here. The birds are here, and the people are here. Of course, the mosquitoes are here and will always be here."

The viral plume soon stretched all the way south to Marathon, in the Florida Keys, when a vacationing seventy-three-year-old woman from Sarasota, suffering from confusion, swollen lymph glands, headache, and a high fever, was diagnosed with West Nile fever. She recovered and was released from the hospital several days after being admitted. By the end of 2001, the virus had infected people in ten eastern states, from Massachusetts to Florida, over an area of half a million square miles, and it had been detected in birds across the eastern half of the United States.

No one expected the virus to stop there, not least of all David Rogers, a professor of ecology at Oxford University who had been tracking the virus since its arrival in the New York City area. Working with colleagues from the National

Aeronautics and Space Administration (NASA), Rogers developed risk maps to predict where the virus was likely to strike next—thus potentially alerting people in the disease path to take precautions. After the Florida outbreak, Rogers began feeding into a computer at Oxford satellite images of ground vegetation, temperature, and other information suggestive of good mosquito habitat. On this data he then superimposed the coordinates of areas in which infected birds had been found. By early 2002, the NASA team had identified Louisiana as a potential trouble spot. True to its prediction, by late summer the epidemic had struck there. Fifty-eight people fell ill. West Nile had also arrived in Mississippi and several nearby states.

The virus struck Louisiana with such fierceness that some speculated it might have mutated into something more virulent than the strain from New York. For one thing, the Louisiana outbreak seemed to be striking a higher percentage of young people than had West Nile outbreaks in the previous three years. During the virus's first two years in the United States, the average age of patients was about sixty-six years; in 2001 it was even higher, seventy. But during the initial 2002 outbreak in the Gulf of Mexico region, the victims' average age was in the upper fifties. Of the fifty-eight cases, twelve of the victims were between forty-five and fifty-nine years of age, and nineteen were younger still. Was the shift in age a coincidence, or had the virus undergone an ominous mutation that gave it the power to overwhelm the relatively healthier immune systems of the young? Time would tell.

"The peculiarity about West Nile virus," David Rogers told me, "is that it appears to be supported by at least thirty species of mosquito vectors in the United States and at least eighty species of bird hosts, not to mention some other animals. Normally diseases—even viral ones—have fewer hosts and vectors. Even the relatively close cousins of West Nile virus, such as yellow fever, have far fewer. The greater the number of vectors and hosts, the more likely a disease is to spread within a new continent. In three years West Nile virus in the US has gone from zero to thirty-four states. That's a record by any standard."

Rogers pointed out that in 2002 alone there were 4,161 documented cases, with 284 deaths, in forty-four states and the District of Columbia. The Midwest was particularly hard hit, with nearly 2,000 documented cases in just Illinois, Michigan, and Ohio. "This must make West Nile fever one of the most important vector-borne diseases in the entire United States—if not the most important—all within three years of its first appearance in New York," he said.

"One expects viruses to travel," Rogers concluded. "They always have, especially when migrating animals are part of the equation. But the rapidity of the spread of West Nile virus is unprecedented. One cannot say exactly where it is going to stop."

8.

It didn't. By the end of 2003, human cases had been reported in all of the lower forty-eight states except for

Maine, Washington, and Oregon. By 2004 it had invaded Oregon; in 2006 it reached Washington State. In 2012 the first human infection in Maine was documented—a thirty-four-year-old man from Cumberland County. As of April 2013, only Alaska and Hawaii had been spared infection with the virus.

Within a decade of its appearance in the United States, West Nile virus had infected nearly 2 million people, causing illness in 360,000 and encephalitis or meningitis in almost 13,000, and killing more than 1,300. The virus forced the implementation of a costly national blood donor screening program to keep the nation's blood supply safe. In the same way that Lyme disease greatly affected when, where—and if—people hiked or went outdoors in the Northeast, West Nile virus discouraged many people from spending time outdoors during mosquito season in many areas. Like Lyme disease, West Nile virus has changed the way of life for many. And that's just the human toll. The virus has killed millions of birds and cut the populations of some species in half. Some have recovered, but others have not.

The bird-borne virus quickly spread along the Atlantic, Mississippi, Central, and Pacific Flyways to southern Canada, Central and South America, and the Caribbean. Probably assisted by the wild-bird trade (legal or illegal), shipping, and air travel, the virus soon infected large areas of Europe and Africa, as well as Australia. Carried by rodents, bats, cats, dogs, horses, ungulates, and reptiles—not to mention humans—it was never at rest for long. By 2013 it had been isolated from more than sixty species of mosquitoes and more than three hundred species of birds.

Warmer temperatures, higher humidity, and heavy rain have all increased human infections with West Nile virus, according to recent research. The largest outbreak since 1999 occurred in the summer of 2012, a year of scorching heat and drought and the warmest year ever in the United States, according to the National Oceanic and Atmospheric Administration. In many ways it mirrored what had happened in 1999, when Enrico Gabrielli got infected. But in 2012 an even more dramatic string of temperature records were broken across the United States.

On June 27, Hill City, Kansas, reached a high of 115 degrees; Indianapolis reached 104, and St. Louis, Missouri, 108 degrees. As if a terrible outbreak of West Nile virus weren't enough, the blazing temperatures and weather that year would also bring floods, wildfires, and storms, with record storm surges to the coasts of New Jersey and New York.

In fact, the 2012 outbreak of West Nile virus would go down as the worst since 2003. All forty-eight lower states had infections in birds, people, or mosquitoes. By year's end there had been more than 2,800 infections in people, with half ending up in meningitis or encephalitis. There were 286 deaths.

Hardest hit had been Texas, which alone had more than 1,700 cases and 76 deaths. During the height of the outbreak, the mayor of Dallas declared a state of emergency and tried to stem the epidemic by authorizing the aerial spraying of a pesticide in the city for the first time since 1966.

Lyle Petersen, director of the CDC's Division of Vec-
tor-Borne Infectious Diseases, stated that in the United
States and other countries, "hot weather seems to promote
West Nile virus outbreaks. And most—many major West
Nile virus outbreaks in Europe, in Africa, and now in the
United States—have occurred during periods of abnormally
hot weather."

Studies have shown that hot weather can increase the
chances of infection in several ways. First, the virus spreads
more quickly in hot weather than in cooler temperatures.
Second, mosquitoes pick up the virus more easily from infected birds in hot weather. Finally, the higher the temperature outside, the more likely the infected mosquito is to pass
the virus to a person.

Whatever role temperatures may have played, the American robin, a thrush with the scientific name *Turdus migra-
torius*, likely played a critical one. Although the robin has
long been revered as a symbol of compassion, joy, and good
fortune, its days as a symbol of happiness and hope may be
numbered. It has become suspect number one in the spread
of West Nile virus.

The robin is the favorite host for the feeding of the main
mosquitoes that harbor and spread the virus. Despite making up a maximum of 20 percent of the avian communities
looked at in a 2011 study, robins were fed on by up to 80
percent of the mosquitoes, earning it the designation of "amplification host." Over the past twenty years, urbanization
has helped to increase the yard-loving robin by as much as

100 percent in some areas. This has led some researchers to conclude that suburbia and one of its proudest symbols— the robin—have likely played right into the hands of West Nile virus.

In fact, both urbanization and climate change are key elements in the emergence of new disease. The recent swine flu pandemic and potentially pandemic circulating bird flus have greatly benefited from domesticated hogs and poultry. Lyme disease increases with forest fragmentation and the increase in human-tolerant species such as mice and deer. The spread of hantavirus increased with the abundance of the mice that carry it. Elsewhere, domestic dogs maintain the transmission of rabies on the Serengeti Plain. Yellow and dengue fever are transmitted by mosquitoes that thrive in populated areas.

With each new or emerging disease, the connections to human activity are becoming better understood. In 2010 the increasing role of humans in fostering diseases such as West Nile virus prompted the CDC to begin offering for the first time direct grants to states and cities to study the health effects of climate change. Currently there are initiatives in seventeen states. Although such actions may help to soften the blow of some new diseases, many others, like West Nile virus, will have established themselves long before we even know they have arrived.

Birds, Pigs, and People:
The Rise of Pandemic Flus

1.

In early April 2009, a ten-year-old boy in San Diego County, California, showed up at an outpatient clinic with fever, severe cough, and vomiting. Suspecting flu, the health worker took a routine throat swab and sent it to the county laboratory.

The lab technician there confirmed influenza A but couldn't determine the subtype, or the "HN," of the virus. (The lab had reagents on hand to determine H3N2 and other common subtypes but not unusual ones.) "When we got a sample that we couldn't subtype, and that doesn't happen very often, we got concerned," said Anna Liza Manlutac, the supervisory microbiologist at the time. She sent the sample to the state lab in Sacramento—which forwarded it to the Centers for Disease Control and Prevention in Atlanta, Georgia—and then awaited the results. Meanwhile, an eight-year-old girl living in Imperial, California, more than 125 miles away, came down with similar symptoms.

On April 15, the CDC identified the virus from the first victim. It wasn't H3N2 or any other common seasonal flu variety but H1N1, or swine flu. This particular H1N1 strain had never before been seen in humans. That meant a potential pandemic. The CDC called it S-OIV, for "swine origin influenza virus," and began preparing for the worst.

Unlike pandemic flus, seasonal flu varieties bear some similarity year to year. Exposure one season leaves people with some immunity to the derivative strains the next. This tends to blunt the impact. But pandemic strains are new, so people carry little or no residual immunity. Pandemic strains can therefore spread faster and make people sicker than seasonal flu. If a pandemic strain happens to be highly virulent, it can fell young and old, healthy and infirm alike and kill far more than the hundreds of thousands seasonal flu kills each year around the globe.

The CDC soon determined that the boy in San Diego County and the girl in Imperial, 125 miles apart, had the same virus. Since they hadn't been infected by a common source, the virus must have been spreading from person to person. Two requirements for a pandemic had been met: a new virus and a contagious one. The question was whether it could sustain large outbreaks and spread worldwide. As for the first two victims, both recovered.

On April 12 Mexican authorities reported an outbreak of severe respiratory disease in the state of Veracruz. (In fact, it turned out that more than 600 cases actually stretched back to early March, and the illness was soon to strike nearly one-third of the population of La Gloria.) Clusters also appeared in Mexico City and San Luis Potosí. Many victims were young adults, who usually are spared the worst effects of seasonal influenza but often are hit hard by pandemic strains. Many were hospitalized. When it turned out that SOIV was behind the outbreaks in Mexico and the United States, the CDC issued a travel health warning recommending that United States travelers postpone nonessential travel to Mexico. By that time, the virus had long since jumped the border.

Like the roots of invasive bamboo, H1N1 was sending up shoots in far-apart places. New cases emerged in Guadalupe County, near San Antonio, Texas; around Houston; and in Ohio and New York. By the end of April, New Zealand and Spain had reported cases. Secretary of Homeland Security Janet Napolitano proclaimed a public health emergency but likened the situation to predicting a hurricane: "The hurricane might not actually hit."

It did. With the virus soon spreading to several countries, the World Health Organization raised the pandemic alert to Phase 4, meaning that the virus was igniting community-wide outbreaks and spreading globally. By the end of April, the CDC had confirmed cases on five continents. Still, WHO hesitated to elevate the alert to Phase 5—signifying an imminent pandemic—because influenza is so unpredictable. History is littered with lessons about crying "pandemic" too soon.

2.

In 1976 a mysterious respiratory outbreak hit Fort Dix, New Jersey, striking army recruit David Lewis, who was returning to town when he developed trouble breathing and collapsed. His commanding officer revived him, and Lewis was taken to the hospital on the base. He was pronounced dead on arrival. Lewis's death, along with the mounting victim toll, set off panic. Fearing a second coming of the 1918–1919 Spanish flu pandemic, the director of the CDC sent a memorandum to the US Department of Health and Human Services urgently recommending mass immunization. President Gerald Ford soon announced a crash program to "inoculate every man, woman and child in the United States." Five months after the initial outbreak at Fort Dix, there was still no evidence of spread beyond the area. Not until seven months after the outbreak was the first vaccine administered. More than 45 million people eventually received it. But the feared pandemic never came.

Sadly, more than five hundred people who had been vaccinated suffered a paralyzing nerve condition, according to research around the time. More than thirty died before the crash vaccination program was suspended. The influenza outbreak's final confirmed toll was 230 cases, with 13 hospitalizations and 1 death. The cure, it was widely believed, had been worse than the disease. (More than three decades later, an analysis using more modern tools called into question whether the vaccines had actually caused the paralysis.)

In 1997, almost twenty-five years after the Fort Dix deba-
cle, came the second pandemic scare. That year, eighteen
people caught a deadly type of flu from live infected poultry
in Hong Kong. Half of them died. But in contrast with the
situation at Fort Dix, "bird flu" didn't fade away. Warnings
abounded that the second coming of Spanish influenza was
really imminent this time. Fifteen years later, deadly bird flu
continues to occasionally infect people around the world.
Yet the dangerous H5N1 has not evolved into a pandemic—
at least not yet.

If public health organizations hesitated to cry pandemic
in 2009, it's because they had learned from 1976 and
1997. But on April 29, 2009, with overwhelming evidence
that the new H1N1 virus was about to explode, the World
Health Organization raised the alert to Phase 5. A pandemic was imminent.

Whatever skepticism past flu scares had spawned, the vig-
ilance had led to vastly improved flu surveillance. By 2009
more than 130 national influenza centers in 101 countries
were conducting year-round surveillance for the appearance
or spread of any new strains. After collection and initial analysis, the samples were sent to one of five World Health Organization (WHO) Collaborating Centers for Reference and
Research on Influenza in Atlanta, Georgia; London; Mel-
bourne, Australia; Tokyo; and Beijing. With data on the new
virus pouring into databases accessible to scientists around
the world, within weeks—sometimes even days—scientific
reports and papers, facilitated by online peer review and

. . . .

publication, were appearing on sites such as Plos.org and updates on Nature.com. Real-time multicolored maps and charts were rapidly unfolding at flu.net and other database sites. The traditional plodding process of scientific review and publication had reached unprecedented speed—just what was necessary to keep up with fast-moving influenza. Yet the virus was leaving its high-tech trackers and scientific sleuths in the dust.

In the past, a pandemic strain took six to nine months to spread around the world, but H1N1 spread throughout the world from Mexico in a matter of weeks. This isn't surprising, given that during a typical flu season more than 2 million people typically fly from Mexico to more than a thousand destinations in over 160 countries. Eighty percent land in the United States or Canada, while others go to Central and South America and the Caribbean Islands. Nearly 10 percent fly to western Europe. And wherever travelers from Mexico went, H1N1 often followed. "Of the 20 countries worldwide with the highest volumes of international passengers arriving from Mexico, 16 had confirmed importations associated with travel to Mexico," according to a 2009 study in the *New England Journal of Medicine*.

Although many people were infected, not until late April did the first US victim of the disease die. Earlier that month, two-year-old Miguel Tejada Vazquez and his mother, vacationing from Mexico, had visited Houston's Galleria mall. (Miguel was the grandson of Mario Vazquez Ráña, a press baron from Mexico and former owner of the United Press International news service, who at the time owned forty-one newspapers in Mexico.) A few days later, Miguel came down with a severe respiratory illness and was rushed to Texas Children's Hospital. He died on April 27.

By May 1 Texas had twenty-eight confirmed cases, second only to New York, where the virus hit St. Francis Preparatory School in Fresh Meadows, Queens. Sixty-nine students became ill. More than thirty New York City schools soon closed. Of the city's 200 confirmed cases at the time, most involved only mild illness. Fifty-five-year-old Mitchell Wiener, an assistant principal at Intermediate School 238 in Hollis, Queens, was not one of them. He died within days of becoming infected.

The CDC's global map, meanwhile, was lighting up with new cases in Germany and Austria, in parts of Asia, and in eighteen other countries. The hundreds of confirmed cases suddenly exploded into 8,500, with 72 deaths, in thirty-nine countries. By the second week in June—barely a month after the ten-year-old boy became ill—there were 30,000 cases in 74 countries.

On June 11, WHO Director-General Dr. Margaret Chan called a news conference and declared that a pandemic was officially under way. It was the first in almost forty years—since the Hong Kong flu of 1967–1968. "The virus is entirely new," she said. "Further spread is considered inevitable. We are all in this together, and we will all get through this, together."

. . . .

3.

"The virus that has caused these infections is actually very interesting," Dr. Nancy Cox, director of the CDC Influenza Division, said at an earlier press briefing. The virus had gene segments from bird, human, and pig flu viruses—what scientists call a "triple reassortant." The pig component had come from both North American swine influenzas and one from Eurasia. Although Dr. Cox found the new virus interesting, she probably hadn't found it completely surprising.

For more than eighty years, a familiar strain of H1N1 had been coursing through American hog farms, causing periodic outbreaks of "swine influenza" in the animals and occasionally minor infections in people who had been in contact with them. Around 1998 this familiar H1N1 virus underwent a major genetic change and emerged in pigs as a triple reassortant. This new three-headed virus—part pig, part bird, part human—caused a rash of outbreaks in North American swine. Then it continued to silently evolve. By 2005 descendants of the triple-headed virus had begun spilling more frequently into people who worked with pigs.

Between 2005 and 2009, eleven triple-reassortant swine flu infections were documented in people. The victims had come in direct contact with the pigs through butchering, at fairs, on a farm, or in live animal markets in the Midwest or Texas. Symptoms included fever, cough, headache, and diarrhea. Four of the victims ended up in the hospital, and two had to be put on ventilators. One of those infected was a previously healthy seventeen-year-old boy from Wisconsin who, a week before falling ill, had helped his brother-in-law slaughter pigs. The boy had pulled a slaughtered pig's front legs forward while his brother-in-law gutted the animal. After a few days with a headache, low back pain, and a cough, the boy recovered. The pigs had no apparent illness, but researchers concluded the boy had most likely been infected by inhaling aerosolized secretions from the slaughtered animal's lungs or airway. All eleven victims ultimately survived.

A three-headed virus that could jump to people was still something fairly new, and such a quirk of viral evolution had not come about without a lot of inadvertent human help. Certain conditions had to be met for it to occur—conditions that would probably not have occurred naturally. First, the three types of flu viruses creating it—bird, human, and pig—had to converge on the same host.

Enter the versatile pig.

Pigs are hospitable to both human and bird flu viruses—not to mention hosting their own strains. In a pig, different viruses can co-infect the same cells. Once cohabiting a single cell, the viruses split apart, exchange genetic material, and replicate. The spun-off virus could have parts of all of them. The idea of the pig as a flophouse for influenza viruses isn't new. A century ago, veterinarian J. S. Koen, an inspector with the US Bureau of Animal Industry, pointed out that flu jumps back and forth between people and pigs. What people didn't know at the time was that cells in the airways of a pig have surface features that happen to fit the landing ports of human flu viruses. (This landing port is what the "H" of a viral subtype refers to, as in H1N1.) Swine tracheal cells not only host human flu viruses; they also host the H portion of bird flu viruses. The pig is therefore something of a biological boudoir for the meeting and mating of flu viruses.

If the pathway that human and pig flus travel between the two species is understood, exactly how the avian strains became part of the three-headed virus is much less so. What is known is that birds—especially aquatic ones—are the natural reservoir of all the influenza A viruses. In fact, only from wild waterfowl and seabirds have all the known subtypes of influenza A been isolated. They are a continuing wellhead of dangerous diversity for the seasonal influenzas that infect people and feed pandemics.

The ecological pathways by which avian flu genes regularly move into humans were probably established long ago. For centuries farmers in southern China and along the rice belt of Vietnam, Thailand, and Cambodia have grown rice in a system that uses domesticated ducks to feed on weeds and insect larvae, snails, and other pests in the paddies. When the rice blooms, the ducks are moved. After the harvest, they are welcomed back to feed on any remaining grain. The rice belt is along major migration routes for waterfowl. Drawn to the artificial wetlands, the wild birds, which are natural carriers of influenza A, land in the paddies and shed the virus in their feces. The domesticated poultry feeding in the paddies pick up flu viruses from the wild birds.

In many regions, pigs have been integrated into this system of seeming ecological beauty and efficiency. But the high agricultural yield hides the messy viral cross contamination behind it. In Thailand, some rice paddies are connected to fish ponds, which are "enriched" with pig manure. Nutrient-rich water from the fish ponds is sometimes used to fertilize the rice paddies. All the while, humans frequent the areas and sometimes live adjacent to them. Little wonder southern China and other parts of Asia have become known as global "flu generators," seeding outbreaks of influenza with new genetic material while occasionally propelling the rise of new and potentially deadly pandemic strains.

More recently, the rising affluence of many in China has increased the numbers of poultry and pigs. "No question. Today there is more of what we need for the virus to move from one host to the other," said the CDC's Nancy Cox. "Poultry is relatively inexpensive to grow to supplement needs and desires of the human population. Swine farms have also grown." What's more, the disposal of pigs that inevitably die from disease outbreaks is a growing challenge to farmers. Prohibited in certain cases from burying the animals, farmers throw them into nearby rivers. In March 2013 thousands of dead pigs festered in the major river flowing through Shanghai.

Expanding markets in cities, where animals of every conceivable type are kept live, butchered, or sold among crowds of shoppers, have created major microbiological thoroughfares for cross contamination, evolution, and human infections with flu and other diseases such as SARS. According to a 2009 study in the *Journal of Molecular and Genetic Med-*

icine, "commercial poultry farms, 'wet markets' (where live birds and other animals are sold), backyard poultry farms, commercial and family poultry slaughtering facilities, swine farms, human dietary habits and the global trade in exotic animals have all been implicated in the spread of influenza A viruses. The 'wet markets' of Southeast Asia, where people, pigs, ducks, geese and chickens (and occasionally other animals) are in close proximity pose a particular danger to public health." Traditional agriculture, expanding pig and duck culture spurred by increasing affluence, and the growth in live animal markets may all have helped to generate new strains of influenza.

In theory, any place where wild and domesticated birds intermingle with humans can create a bridge for a flu virus to cross. In 1996 the *Lancet* reported that a housewife in England, who kept a duck house next to a pond frequented by Canada geese, mallards, and other wild birds, appeared to be among the first cases on record in which a particular type of bird flu jumped directly to humans. Fortunately, she came down with only a bad case of conjunctivitis. Such isolated cases have probably occurred for centuries. But the vast scale of live animal agriculture today, especially in China and Southeast Asia, has become central to the maintenance of seasonal flu and the creation of pandemics.

Large hog farms in the United States, Mexico, and elsewhere probably also give viruses ample chance to meet and mix in their favorite host before inflicting the human caretakers. Genetic analysis of the 2009 H1N1 outbreak showed that swine can give rise to pandemics and that factory farm workers' occupational exposure to pigs vastly increases that risk.

It may not have been coincidental that five-year-old Edgar Hernandez of La Gloria, Mexico, was one of the earliest confirmed cases of H1N1. La Gloria, ground zero for the 2009 pandemic, is located in a major hog-farming area, not far from the town of Perote, home of the massive hog-farming company Granjas Carroll de México. A later survey of the hog farm failed to turn up the virus, and lack of sampling in the years before the virus emerged would make pinpointing its origins impossible—beyond knowing that it emerged in swine, where it had been circulating undetected for a decade before spinning off triple-headed strains whose descendants ultimately sparked the 2009 epidemic.

Nevertheless, Texas resident Steven Trunnell was convinced that a Mexican hog-farming operation gave rise to the flu virus that killed his wife, Judy. On April 14 Judy, who was eight months pregnant, developed achiness, dry cough, and a slight fever. The next day she visited her obstetrician-gynecologist. A rapid diagnostic test showed that she had the flu. Five days later she went to a local emergency room with a fever and gasping for air. Her lungs were filling with fluid, so she was intubated and put on a respirator. Later that day, a healthy baby daughter was born by an emergency cesarean delivery. But it was too late for Judy, who passed away on May 4—the first American citizen to die from H1N1.

On May 11 Steven, a paramedic, filed a petition in the

district court of Cameron County, Texas, seeking to depose officials of Smithfield Foods, part owner of the hog operation in Perote. The petition claimed that "it is likely that the creation and spread of this lethal strain of swine flu may have been caused, in part, by historically unsanitary conditions which Smithfield Foods knowingly caused to occur in Mexico in connection with the operation of the largest pig farm business in the world." The petition stated that "it is reasonable to expect that this area around La Gloria is 'ground zero' for the H1N1-2009 swine influenza virus." Ultimately, the petition went nowhere.

A year and a half after it began, the worst of the 2009 swine flu pandemic had passed, and on August 10, 2010, the World Health Organization announced that the world had entered the "post-pandemic period." "The new H1N1 virus has largely run its course," WHO said. "This time around, we have been aided by pure good luck. The virus did not mutate during the pandemic to a more lethal form." Although it was far from another "pandemic that never was," it was, by many accounts, far milder than most had predicted.

According to Marc Lipsitch of Harvard School of Public Health, it was probably "the mildest pandemic on record—compared to the three that happened in the 20th century," although it had a disproportionate impact on children and young adults. The CDC's director, Thomas Frieden, on the other hand, said that "any flu season that kills at least three times more children than a usual flu season—I think it would be very misleading to describe that as mild."

By the time it was over, the pandemic had affected over 214 countries and caused more than 18,000 laboratory-confirmed deaths—including more than 250 children. A study in the *Lancet* cautioned that "this number is likely to be only a fraction of the true number of the deaths associated with 2009 pandemic influenza A H1N1." The study concluded that between April 2009 and August 2010, more than 200,000 people had been infected—80 percent of them adults under the age of sixty-five. More than half of the infections occurred in Southeast Asia and Africa.

Probably never before had so much been learned so quickly about a virus. Three years after H1N1 appeared, a search of medical literature showed that more than 2,500 papers on it had been published, making it among the most studied pandemics in history—and generating an enormous amount of data that would help scientists to better understand future outbreaks.

The good news was that the pandemic had come—and gone—with far less damage than many had anticipated. The bad news was that H1N1 wasn't even the pandemic everyone had been predicting and preparing for. The original suspect—the feared H5N1 bird flu from Asia—had been on the loose since 1997 and was considered far more lethal.

4.

This most feared pandemic—H5N1, or "bird flu"—began in Hong Kong in 1997. Bird flu first came to light when a three-year-old boy in Hong Kong was hit with a fever, sore throat,

and cough. He was admitted to the hospital's pediatric intensive care unit, where he soon died of severe respiratory
distress. Like the later H1N1, bird flu had never before been
seen in humans. Of the eighteen people infected in Hong
Kong during the initial outbreak, six died.

Today, fifteen years later, bird flu is still traveling, having
spread widely in poultry. And where poultry outbreaks began,
human infections weren't long in following. In almost all human infections the virus jumped directly from birds to people, but there may be some limited person-to-person transmission, according to the CDC. Between 1997 and 2013,
more than 600 people were infected. More than half died.
The situation harkens back to the days when H1N1 also had
limited human transmission. But fortunately, when it comes
to influenza the past does not always predict the future.

If the "sudden" appearance of swine flu in 2009 misled
the public to believe that pandemics seem to explode out
of nowhere, the lingering bird flu from 1997 taught that,
in reality, they almost never do. And if H1N1 tricks us into
believing that influenza is a medical issue best addressed by
physicians, H5N1 tells us that the real cause is ecological.

In some ways the ecological trajectories of swine flu and
bird flu were similar. But whereas H1N1 was mostly at home
in pigs, H5N1 has been most at home in birds, which gave it
birth. The year before the 1997 Hong Kong outbreak, H5N1
was isolated from a flock of sick geese in Guangdong Province, China. No one can say exactly how the virus first arose,
only that it had probably been carried there by birds. But the

virus may already have been widely dispersed in wildfowl, and it may have been only by chance that it was detected in Guangdong instead of elsewhere.

H5N1 is a type of poultry disease called highly pathogenic avian influenza, or HPAI. This flu virus and other flu viruses that infect birds have caused among the largest outbreaks of animal disease ever recorded, with several hundred million wild birds, geese, chickens, turkeys, and ducks having died from it. Long before the flu virus was identified in these massive outbreaks, the disease it caused was known as "fowl plague," first identified in Italy in 1878. The virus was spread through the transport of fowl to poultry exhibitions and shows in Europe in the late 1800s and early 1900s. But not until 1955 was it determined to be an influenza A virus. Although avian influenza is now controlled in the United States, in 1983 and 1984 an outbreak in the northeastern United States led to the destruction of more than 17 million birds and cost $65 million, causing the retail price of eggs to jump by nearly 30 percent at the time. A 2004 outbreak in Canada led to several hundred million dollars in losses.

Although mostly limited to birds, avian influenzas such as HPAI have long been known to cause occasional mild illness in humans. Their versatility in infecting other species is also well documented. In 1986 two different flu viruses from gulls were found to have infected a pilot whale. Avian influenza viruses have also caused periodic die-offs of seals near New England.

Not long after the 1997 H5N1 outbreak in Hong Kong,

the virus began spreading to mainland China and elsewhere in Southeast Asia, sowing occasional human infections along the way. Then, in 2003, new human H5N1 infections emerged in Vietnam, followed by sporadic cases in Europe, Africa, and the Middle East. Reports of human infections in Thailand soon followed. The virus was also killing 100 percent of the poultry it infected there and in Vietnam. Wherever there were human cases, they seemed to follow outbreaks in poultry.

Although spread in part by migrating birds, the virus had become so entrenched in poultry throughout regions of China that distinct regional variations of the bug evolved. Detached from wild birds, where it arose, it is now mapping its own evolutionary course. The movement of poultry has reintroduced these regional variations from one region or country to the next, adding further momentum.

Bird flu's forays into other species didn't stop, and it seemed to run rampant across interspecies borders. In late 2003 cases of H5N1 were reported in dogs, cats, pigs, and weasels. In December of that year, two tigers and two leopards in a Thai zoo died after being fed carcasses of slaughtered chickens that had been having respiratory problems. At about the same time the tigers and leopards died, there was an outbreak of H5N1 on nineteen poultry farms in Korea. Just before the outbreak, the owner had on many occasions seen magpies entering an area where chicken feces were disposed of. Dead magpies later found on the farm had been infected by the virus.

In Japan, dead crows found near chicken pens had also been infected with the same virus originally found in Guangdong Province, or one closely related to it. The virus decimated poultry wherever it struck. Over 100 million domesticated birds were culled in a futile effort to contain the first wave of the virus. A second wave swept through poultry in China, Indonesia, Thailand, and Vietnam in late 2004. And another outbreak at a Thai zoo led to the deaths of 147 tigers.

By 2005 a third wave of H5N1 was sweeping through Southeast Asia. Human cases were also being reported every month in Asia, eastern Europe, Africa, and the Far East. Still, there was little, if any, transmission from one person to the next. In October 2005, 276 smuggled songbirds died en route to Taiwan from mainland China. Later tested for disease, the birds were found to be infected with H5N1. It wasn't the first time that contraband could have contributed to the spread. The year before, two eagles hidden in tubes and smuggled into Brussels from Thailand were found to be infected. Although showing no symptoms, both birds were euthanized.

In 2005 three civet cats died of H5N1 in Vietnam. Often bought and sold in street markets in China, civet cats were double-crossed: they had also spread SARS—severe acute respiratory syndrome—to people during that outbreak in 2003.

In April 2005, H5N1 killed more than six thousand bar-headed geese, gulls, shelducks, cormorants, and other wild birds at Qinghai Lake in central China—the first time the virus had shown sustained transmission in waterfowl. It had probably been carried there by wild birds that picked it up from poultry in southern China.

After the Qinghai Lake outbreak, dead migratory birds were found in western Siberia and in Kazakhstan and Tibet, as well as in Mongolia. H5N1 continued to spread among poultry in Turkey and Romania and in mute swans in Croatia and Hungary. Birds fell ill from H5N1 at a zoo in Jakarta, a dead flamingo was found in Kuwait, and dead swans were found in Iraq and Egypt; in Bulgaria, Greece, Italy, Austria, Hungary, Germany, Slovakia, Poland, and Denmark; and elsewhere in Europe. Poultry were infected in Afghanistan, Pakistan, Jordan, and Israel and in several other North African countries. By 2006 human cases were beginning to surge again, with most of them in Indonesia and Egypt. Sporadic human infections continue to this day.

In 2006 influenza researcher Robert Webster declared that "the likelihood of an H5N1 influenza pandemic seems high, and the consequences could be catastrophic. Recent findings suggest that the 1918 'Spanish flu' pandemic may have resulted from a similar interspecies transmission event in which a purely avian virus adapted directly to human-to-human transmission."

Several years after Webster's warning, studies suggested that the virus could be only a few simple mutations away from contagiousness. In one study, scientists inserted into the 2009 swine flu virus a mutant version of a key viral protein from bird flu. A mere four mutations later, the hybrid strain was able to strongly bind to mammalian cells and replicate enough to saturate respiratory droplets—the beginnings of aerosol transmission—that is, it developed the ability to spread easily from one person to the next. Technical details of the research were so worrisome to the terrorism-sensitive US National Science Advisory Board for Biosecurity that it recommended researchers withhold key details when publishing their work. After a heated six-month debate between government officials and scientists, two papers detailing the results were finally published in their entirety in *Nature* and *Science*.

Even as public health officials at the World Health Organization and the Centers for Disease Control and Prevention wrote swine flu into the history book, the catastrophic potential of bird flu weighed heavily on their minds. One might have hoped that pandemic scares would end there. They didn't. In 2012 yet a third contender for the title of Next Pandemic arose.

5.

The third "pandemic" begins in early April 2013 when a sixty-year-old woman from Zhejiang Province, China, is hospitalized, barely able to breathe. Fourteen other cases of severe respiratory difficulty quickly come to light in Zhejiang Province, Shanghai, and Anhui Province. All of the victims are hospitalized. Six die. The virus identified, H7N9, has never been seen before in humans. Because it is new to people, it has "potential pandemic" written in its H surface protein.

Setting H7N9 ominously apart from even H5N1 are genetic changes, according to authors of a 2013 study, that "probably facilitate binding to human-type receptors and efficient replication in mammals . . . highlighting the pandemic potential."

With H5N1, H1N1, and now H7N9—all new to human experience—the earth seems to be passing through a meteor shower of new flu viruses. H1N1 has already struck, bird flu appears to be a near miss, and H7N9 is passing somewhere between. The genetic makeup of H7N9 suggests it probably originated from Eurasian avian influenza viruses. Other components echoed duck, chicken, and even pigeon ancestry.

So the story goes.

The cases are initially limited to Shanghai and neighboring regions. By the third week in April two people are infected in Beijing, to the north, and two more in Henan Province—with live poultry shipped from Shanghai the likely source of the spread. But with some 6 billion domesticated birds shipped annually throughout China, the viral trail is almost impossible to follow. There's still no evidence of the virus spreading widely among people, but the scattered reports are worrisome. Jeremy Farrar, director of the Oxford University Clinical Research Unit in Ho Chi Minh City, Vietnam, told the journal *Nature*, "I think we need to be very, very concerned."

Because H5N1 leaves a trail of dead poultry, the virus's whereabouts are known. But H7N9 can silently infect birds, flying under the radar, spreading through flocks undetected and in turn infecting people in poultry markets, far from where the last human cases were seen. By mid-April there are 63 infections and 14 reported deaths—up from 24 cases in a single week. Within the first two weeks of its appearance, H7N9 virus is infecting more people than H5N1 has since 1997.

In April 2013, the first asymptomatic case of H7N9 is detected in humans—a four-year-old girl who had been in contact with a seven-year-old who fell seriously ill. The occurrence of silent cases among humans, as among poultry, means the virus may be evading surveillance. US Secretary of Health and Human Services Kathleen Sebelius declares that the avian influenza virus has a "significant potential to affect national security."

Shortly after H7N9's emergence, studies show it has already mutated in people since jumping from birds—a mutation that allows the virus to grow well at a temperature similar to that of the human upper respiratory tract.

By August 2013 it has infected more than 130 people in China, with more than 40 deaths. As with its dangerous cousin H5N1, which continues to evolve, no one can say where the new virus will end.

The rest remains to be seen.

··· Epilogue ···

MERS-CoV and Beyond

It is only a matter of time, many epidemiologists warn, until another epidemic on the scale of the Spanish influenza outbreak of 1918–1919, or the current HIV/AIDS pandemic, sweeps across the globe.

More than a decade ago, the National Academy of Sciences' Institute of Medicine cautioned:

> Today's outlook with regard to microbial threats to
> health is bleak on a number of fronts. . . . Pathogens—
> old and new—have ingenious ways of adapting to and
> breaching our armamentarium of defenses. We must
> also understand that factors in society, the environment,
> and our global interconnectedness actually increase
> the likelihood of the ongoing emergence and spread of
> infectious diseases.

Since 9/11, infectious disease threats have increasingly

assumed a mantle of bioterrorism. As Jane Evans of the Department of Military Strategic Studies, US Air Force Academy, wrote bluntly in *Global Security Studies* in 2010, "infectious diseases threaten national security."

A number of potential infectious disease threats have emerged, but in most cases they have either subsided or failed to materialize. In the late 1990s there was bird flu, or H5N1. Although still present, it has not become the pandemic that many have feared. Not yet.

In 2009 swine flu emerged, and though it became a pandemic, its impact was less than many had predicted.

In 2013 a new form of lethal bird flu emerged, H7N9. Like H5N1, it lingers in the background, its utter unpredictability preventing the scientific community from making any meaningful assessment of its true threat.

And just before the latest bird flu emerged, a new virus was isolated from a patient in Jeddah, Saudi Arabia, who had come down with a severe and sudden case of pneumonia and renal failure. A short time later that summer, in 2012, the same virus was isolated from a patient in London with severe respiratory illness. That patient had been in the Middle East and had recently returned to Great Britain. Ultimately the cases were connected to an even earlier outbreak, when health care workers at an intensive care unit of a hospital in Jordan came down with acute symptoms. Two of them died. All of the cases turned out to have been caused by a newly discovered coronavirus.

The virus was later named Middle East respiratory syndrome coronavirus, or MERS-CoV. Although different from the SARS virus of 2003–2004, it is a coronavirus of the same family and causes SARS-like symptoms, including fever, cough, and shortness of breath—all associated with severe acute respiratory illness. The SARS virus had originated in China, had infected 8,000 people and killed about 10 percent of those infected, and was traced back to civet cats.

MERS-CoV has so far been much deadlier—if slower to spread. By May 9, 2013, 33 laboratory-confirmed cases had been reported—24 from Saudi Arabia, 2 from Qatar, 2 from Jordan, 3 from the United Kingdom, 1 from the United Arab Emirates, and 1 from France. Eighteen were fatal.

Most worrisome was the World Health Organization's announcement on May 12, 2013, that the virus was spreading between people in close contact. By July it had infected 85 people, including several in the United Kingdom, Italy, and France. More than half had died. But many milder cases have probably occurred undetected.

Scientists don't know the natural reservoir of the virus, although evidence of genetically similar ones has recently been isolated from camels and bats.

As of September 2013, 130 people had been infected, with 58 deaths. The vast majority of the cases were in Saudi Arabia. With cases only sporadically emerging, in July the World Health Organization met and declared "that the current MERS-CoV situation is serious and of great concern, but does not constitute a [public health emergency] at this time."

Once again, a threat emerges. As our attention is—understandably—diverted by these frightening new illnesses, old threats, such as antibiotic-resistant disease, which kills some 20,000 people every year in the United States, continue to grow. Some of the gravest threats remain the subtle genetic changes we foster in old scourges. Our attention may be focused on the newest epidemic, but death is often in the details of the old ones.

When a frightening new disease does come along, only a deep sense of bias and denial permits us to point reflexively at a phenomenon like bioterrorism as the cause while ignoring our own collective promulgation of the global environmental disruption that has given rise to so many new infectious diseases. We frequently spell out the dangers of deliberate genetic engineering even as we maintain what amount to giant genetic engineering laboratories—in the form of the intensive agricultural systems that create agents such as the one responsible for mad cow disease, which terrorize in their own way.

But even in an age of mounting epidemics, there is hope. With early detection, many epidemics will be contained. For those that are not, medical technology may offer some respite. Changes in personal behavior alone can greatly reduce the risks of many diseases, such as HIV/AIDS. Mad cow disease was quelled by banning the feeding of meat and bone meal to cattle, allowing them to be mostly herbivores again. As the overuse of antibiotics has caused many bacteria to become resistant, more sparing use of antibiotics may help roll back the tide of some untreatable bacterial infections.

But what of the deep ecological, demographic, and industrial roots of the surge in new epidemics? Will humans be able to ameliorate the upset predator-prey balances that have helped precipitate Lyme disease? With the urgent need for dietary protein in some parts of Africa, can people reduce the massive consumption of bushmeat that may be predisposing humans to new forms of HIV? Will societies be able to curb the effects of climate change that are causing some disease-causing organisms to spread to new areas or otherwise proliferate? Although we are learning much about the ecological origins of new diseases, it remains to be seen whether we will address these mounting epidemics at their roots. This will require more than devoting ourselves to new treatments and cures. It will also require curing the cause—and that means protecting and better addressing the ecological integrity upon which our health, and that of so many other species, often depends.

Notes

Introduction

xiii *strange new disease in the city:* J. Steinhauer, "As Fears Rise about Virus, the Answers Are Elusive," *New York Times*, September 29, 1999.

xiii *in some cases were comatose. Several soon died:* D. S. Asnis et al., "The West Nile Virus Outbreak of 1999 in New York: The Flushing Hospital Experience," *Clinical Infectious Diseases* 30, no. 3 (2000): 413–418.

xiii *Black Death, which wiped out as much as one-third of Europe's population in the 1300s, or . . . Spanish influenza:* W. H. McNeill, *Plagues and Peoples* (Garden City, NY: Anchor Press/ Doubleday, 1976).

xiv *severe acute respiratory syndrome—later known as SARS:* World Health Organization, "Severe Acute Respiratory Syndrome," *Weekly Epidemiological Record* 78, no. 12 (2003): 81–88; World Health Organization, "WHO Issues a Global Alert about Cases of Atypical Pneumonia: Cases of Severe

. . . .

Respiratory Illness May Spread to Hospital Staff," press release, March 12, 2003; L. K. Altman and K. Bradsher, "Official Warns of Spread of Respiratory Disease," *New York Times*, March 30, 2003; World Health Organization, Communicable Disease Surveillance 157 and Response, "One Month into the Global SARS Outbreak: Status of the Outbreak and Lessons for the Immediate Future," Update 27, April 11, 2003.

xv   *Liu Jianlun, a sixty-four-year-old kidney specialist:* K. W. Tsang et al., "A Cluster of Cases of Severe Acute Respiratory Syndrome in Hong Kong," *New England Journal of Medicine*, early online publication, April 2003; E. Rosenthal, "From China's Provinces, a Crafty Germ Breaks Out," *New York Times*, April 27, 2003; D. G. McNeil Jr., "Disease's Pioneer Is Mourned as a Victim," *New York Times*, April 8, 2003.

xv   *It had begun to invade Beijing and other cities. It had also arrived in the United States:* E. Hitt, "Early Inklings about SARS," *Scientist*, March 24, 2003.

xv   *flew from Hong Kong to Frankfurt and Munich . . . before even suspecting he had contracted a new disease:* K. Bradsher, "Carrier of SARS Made Seven Flights before Treatment," *New York Times*, March 11, 2003.

xvi   *"a worldwide health threat":* World Health Organization, "WHO Issues a Global Alert about Cases of Atypical Pneumonia: Cases of Severe Respiratory Illness May Spread to Hospital Staff," press release, March 12, 2003.

xvi   *in China, where dozens of new cases were reported every day:* World Health Organization, Communicable Disease Surveillance & Response (CSR), "Cumulative Number of Reported Probable Cases of Severe Acute Respiratory Syndrome (SARS): From 1 November 2002 to 30 April 2003," April 30, 2003.

xvi   *fatality rate, at least in Hong Kong, was closer to 15 percent:* C.

A. Donnelly et al., "Epidemiological Determinants of Spread of Causal Agent of Severe Acute Respiratory Syndrome in Hong Kong," *Lancet*, early online publication, May 7, 2003.

xvi   *suggested that the SARS virus had come from a nonhuman animal:* British Columbia Cancer Agency, "2003 News— 2003/04/12: Genome Sciences Centre Sequences SARS Associated Corona Virus," April 4, 2003; Centers for Disease Control and Prevention, "SARS: Genetic Sequencing of Coronavirus," CDC Telebriefing Transcript, April 14, 2003.

xvii *Nearly 75 percent of new human diseases . . . are carried by wild or domestic animals:* P. Daszak, personal communication, April 2003; P. Daszak, A. A. Cunningham, and A. D. Hyatt, "Emerging Infectious Diseases of Wildlife: Threats to Biodiversity and Human Health," *Science* 287, no. 5452 (2000): 443–449.

xvii *smallpox from cattle . . . the common cold from horses:* T. McMichael, *Human Frontiers, Environments, and Disease: Past Patterns, Uncertain Futures* (Cambridge, England: Cambridge University Press, 2001).

xvii *almost no way to eradicate it:* G. Kolata, "Now That SARS Has Arrived, Will It Ever Leave?" *New York Times*, April 27, 2003. Virologist and Nobel laureate Frederick C. Robbins is quoted as saying, "If you have an animal reservoir, unless you eradicate it in the animal, you can't eradicate it."

xvii *monkeys infected with Ebola virus were imported into Virginia:* Centers for Disease Control and Prevention, "Ebola Virus Infection in Imported Primates—Virginia, 1989," *Morbidity and Mortality Weekly Report* 38, no. 48 (1989): 831–832, 837–838.

xviii *"close the book on infectious diseases":* McMichael, *Human Frontiers*, 88.

xviii *"diseases that seemed to be subdued . . . are fighting back with*

. . . .

renewed ferocity": World Health Organization, *The World Health Report 1996: Fighting Disease, Fostering Development* (Geneva: World Health Organization, 1996), 1.

xviii *"exacerbate social and political instability in key countries and regions"*: J. C. Gannon, *The Global Infectious Disease Threat and Its Implications for the United States*, NIE 99-17D (Washington, DC: Central Intelligence Agency, 2000), 5.

xviii *residents of Chagugang . . . rioted*: E. Eckholm, "Fear: SARS Is the Spark for a Riot in China," *New York Times*, April 28, 2003.

xix *kills nearly 2 million people annually. Half of the victims are children under five years of age*: P. Martens and L. Hall, "Malaria on the Move: Human Population Movement and Malaria Transmission," *Emerging Infectious Diseases* 6, no. 2 (2000): 103–109.

xix *resistant to chloroquine, a mainstay of malaria treatment*: X. Z. Su et al., "Complex Polymorphisms in an 330 kDa Protein Are Linked to Chloroquine-Resistant *P. falciparum* in Southeast Asia and Africa," *Cell* 91 (1997): 593–603.

xix *malaria had been found in both humans and mosquitoes in an American community*: N. Wade, "Two Cases of Malaria Are Acquired in U.S., a Rarity," *New York Times*, September 7, 2002; "Virginia Mosquitoes Found with Malaria," *New York Times*, September 29, 2002.

xix *increase in malaria has been linked to a warming global climate*: S. I. Hay et al., "Climate Change and the Resurgence of Malaria in the East African Highlands," *Nature* 415, no. 6874 (2002): 905–909.

xix *Maui, Hawai'i, reported its first case of dengue fever in more than fifty years*: Health Canada, Population and Public Health Branch, "Travel Health Advisory: Dengue Fever in Hawaii," April 18, 2002.

xx   *led the Centers for Disease Control and Prevention to create an
     entirely new journal:* D. M. Morens and A. S. Fauci, "Emerg-
     ing Infectious Diseases in 2012: 20 Years after the Institute
     of Medicine Report," *MBio* 3, no. 6 (December 11, 2012),
     doi:10.1128mBio.00494-12.

xx   *The scientists who identified it called it Heartland virus:* L. K.
     McMullan et al., "A New Phlebovirus Associated with Se-
     vere Febrile Illness in Missouri," *New England Journal of
     Medicine* 367 (August 30, 2012): 834–841, doi:10.1056/NEJ
     Moa1203378; D. Grady, "New Virus Tied to Ticks Poses Puz-
     zle for Doctors," *New York Times*, September 3, 2012.

xxi *Plagues are striking a wide range of other species:* Daszak, Cun-
     ningham, and Hyatt, "Emerging Infectious Diseases of Wild-
     life."

xxi *a brain-destroying affliction . . . is spreading among wild deer
     and elk in the western United States:* Ibid.

xxii *we are causing or exacerbating many of these ecodemics:* In
     coining the term "ecodemics" to describe these outbreaks, I
     am not attempting to replace the word "epidemic" but merely
     emphasizing the ecological origins of many new diseases.

xxii *"Show me almost any new infectious disease":* Daszak, personal
     communication, April 2002.

xxii *major, extended waves of epidemics:* McNeill, *Plagues and Peo-
     ples*; McMichael, *Human Frontiers*.

xxii *close human contact with cattle and other livestock gave mi-
     crobes a new bridge for jumping to humans:* McMichael, *Hu-
     man Frontiers*. McMichael writes (p. 101): "Smallpox arose
     via a mutant pox virus from cattle. Measles is thought to have
     come from the virus that causes distemper in dogs . . . the
     common cold from horses, and so on."

xxii *global exploration then ushered in a third phase of epidemics:*
     McMichael, *Human Frontiers*.

. . . .

xxiii *a fourth phase of epidemics:* Ibid.

xxiii *Nipah virus:* World Health Organization, "Nipah Virus Fact Sheet," Fact Sheet no. 262, September 2001; K. B. Chua et al., "Nipah Virus: A Recently Emergent Deadly Paramyxovirus," *Science* 288, no. 5470 (2000): 1432–1435; P. Daszak, A. A. Cunningham, and A. D. Hyatt, "Anthropogenic Environmental Change and the Emergence of Infectious Diseases in Wildlife," *Acta Tropica* 78, no. 2 (2001): 103–116.

xxvi *"Whether we like it or not, we are caught in the food chain":* Quoted in S. S. Morse, ed., *Emerging Viruses* (New York: Oxford University Press, 1993), 36.

Chapter 1. The Dark Side of Progress: Mad Cow Disease

1   *"majestic herds of cattle":* W. Wordsworth, "Memorials of a Tour on the Continent, 1820," no. 35, "After Landing—the Valley of Dover, November 1820," in *The Complete Poetical Works* (London: Macmillan, 1888).

5   *"spongiform encephalopathy" on the necropsy form:* Copy of necropsy form signed by C. Richardson and dated September 19, 1985.

5   *Wells confirmed Richardson's diagnosis:* N. Phillips, J. Bridgeman, and M. Ferguson-Smith, "The Identification of a New Disease in Cattle," chap. 1 in N. Phillips, J. Bridgeman, and M. Ferguson-Smith, *The BSE Inquiry,* vol. 3 (London: House of Commons, 2000), sec. 1.7; Phillips, Bridgeman, and Ferguson-Smith, "Statement No. 69: Carol Richardson," in *BSE Inquiry.*

5   *"a novel progressive spongiform encephalopathy in cattle":* G. A. H. Wells et al., "A Novel Progressive Spongiform Encephalopathy in Cattle," *Veterinary Record* 121 (1987): 419–420.

5   *the first-ever documented case of mad cow disease:* Phillips, Bridgeman, and Ferguson-Smith, "Identification of a New Disease in Cattle."

6   *The first human TSE, Creutzfeldt-Jakob disease*: P. Brown and
    R. Bradley, "1755 and All That: A Historical Primer of Trans-
    missible Spongiform Encephalopathy," *British Medical Jour-
    nal* 317, no. 7174 (1998): 1688–1692.

6   *another human TSE, kuru, was identified in Papua New
    Guinea*: Ibid.

6   *By 1988 more than 2,000 cows had been stricken*: World
    Health Organization, "Bovine Spongiform Encephalopathy
    (BSE)," Fact Sheet no. 113, revised June 2001; Department
    for Environment, Food and Rural Affairs (London), *Bovine
    Spongiform Encephalopathy in Great Britain: A Progress Re-
    port —December 2001* (London: Department for Environ-
    ment, Food and Rural Affairs, 2001).

6   *almost 1,000 new cases . . . every week*: Centers for Disease
    Control and Prevention, "New Variant CJD: Fact Sheet,"
    April 18, 2002.

6   *"vets have no cure"*: D. Brown, "Incurable Disease Wiping Out
    Dairy Cows," *Sunday Telegraph* (London), October 25, 1987.

7   *woman . . . diagnosed with Creutzfeldt-Jakob disease*: Phillips,
    Bridgeman, and Ferguson-Smith, *BSE Inquiry*; "Emergence
    of Variant CJD," chap. 5 in Phillips, Bridgeman, and Fergu-
    son-Smith, *BSE Inquiry*, vol. 8, *Variant CJD (vCJD)*, sec. 5.3.

7   *according to one study*: Phillips, Bridgeman, and Fergu-
    son-Smith, *BSE Inquiry*; "History of CJD Surveillance up to
    1990," chap. 2 in Phillips, Bridgeman, and Ferguson-Smith,
    BSE Inquiry, vol. 8, *Variant CJD (vCJD)*, sec. 2.4.

7   *average age of victims . . . was fifty-seven*: Phillips, Bridgeman,
    and Ferguson-Smith, *BSE Inquiry*; "History of CJD Sur-
    veillance up to 1990," chap. 2 in Phillips, Bridgeman, and
    Ferguson-Smith, *BSE Inquiry*, vol. 8, *Variant CJD (vCJD)*,
    sec. 2.4.

7   *CJD in the young*: Phillips, Bridgeman, and Ferguson-Smith,

. . . .

"Emergence of Variant CJD," chap. 5 in *BSE Inquiry*, vol. 8, *Variant CJD (vCJD)*, sec. 5.28.

7   *Conventional wisdom held:* S. B. Prusiner, "The Prion Diseases," *Scientific American* (January 1995): 48–57.

7   *woman had been associated with a farm:* Phillips, Bridgeman, and Ferguson-Smith, "Emergence of Variant CJD," chap. 5 in *BSE Inquiry*, vol. 8, *Variant CJD (vCJD)*, sec. 5.3.

7   *"the risk of Bovine Spongiform Encephalopathy to humans is remote":* Phillips, Bridgeman, and Ferguson-Smith, "The Southwood Working Party, 1988–1989," chap. 4 in *BSE Inquiry*, vol. 2, *The Link between BSE and vCJD*, sec. 4.1. This conclusion is stated in various ways throughout the *BSE Inquiry*. See also, for example, Phillips, Bridgeman, and Ferguson-Smith, "History of CJD Surveillance up to 1990," chap. 2 in *BSE Inquiry*, vol. 8, *Variant CJD (vCJD)*, sec. 2.26.

7   *a sixty-one-year-old dairy farmer:* Phillips, Bridgeman, and Ferguson-Smith, "Emergence of Variant CJD," chap. 5 in *BSE Inquiry*, vol. 8, *Variant CJD (vCJD)*, sec. 5.7; S. J. Sawcer et al., "Creutzfeldt-Jakob Disease in an Individual Occupationally Exposed to BSE," *Lancet* 341 (1993): 642.

7   *"first report of CJD in an individual with direct occupational contact":* Sawcer et al., "Creutzfeldt-Jakob Disease in an Individual."

8   *dismissed the notion of a link between mad cow disease and CJD:* Quoted in D. Brown, "Farmer Dies of Rare Brain Disease after BSE Hits Herd," *Daily Telegraph* (London), March 9, 1993.

8   *"no evidence . . . of BSE passing from animals to humans":* Quoted in "Special Report: What's Wrong with Our Food? How the Death of Cow 133 Started a Tragic Chain of Events," *Observer* (London), October 1, 2000.

8   *had vouched for the safety of beef*: BBC-TV, "John Gummer and His Daughter," May 1990.

8   *"there is no need for people to be worried"*: Ibid.

8   *There had been three cases of BSE on his farm*: P. T. G. Davies, S. Jahfar, and I. T. Ferguson, "Creutzfeldt-Jakob Disease in Individual Occupationally Exposed to BSE," *Lancet* 342 (1993): 680; Phillips, Bridgeman, and Ferguson-Smith, "Emergence of Variant CJD," chap. 5 in *BSE Inquiry*, vol. 8, *Variant CJD (vCJD)*, secs. 5.16–5.24; J. Hope, "'Mad Cow' Farmer Dies," *Daily Mail* (London), August 12, 1993; K. Perry, "Mad Cow Fear as Second Farmer Dies," *Today* (London), August 12, 1993.

8   *a third dairy farmer*: P. E. M. Smith, M. Z. Zeidler, and J. W. Ironside, "Creutzfeldt-Jakob Disease in a Dairy Farmer," *Lancet* 346 (1995): 898; "Illness Link with Mad Cow Disease?" *Cornishman* (Cornwall), December 1, 1994; Phillips, Bridgeman, and Ferguson-Smith, "Emergence of Variant CJD," chap. 5 in *BSE Inquiry*, vol. 8, *Variant CJD (vCJD)*, secs. 5.16–5.33.

9   *did not require "the Government to revise the measures"*: Phillips, Bridgeman, and Ferguson-Smith, "Emergence of Variant CJD," chap. 4 in *BSE Inquiry*, vol. 11, *Annex 2 to Chapter 4: SEAC Meetings* (see the heading "Special Meeting 13/1/95").

9   *should a fourth case arise, the tide of probability would turn*: Phillips, Bridgeman, and Ferguson-Smith, "Emergence of Variant CJD," chap. 5 in *BSE Inquiry*, vol. 8, *Variant CJD (vCJD)*, sec. 5.39.

9   *a fourth ill farmer came to light*: Ibid., sec. 5.66.

9   *fifteen-year-old . . . came down with CJD*: Phillips, Bridgeman, and Ferguson-Smith, "Emergence of Variant CJD," chap. 5 in *BSE Inquiry*, vol. 8, *Variant CJD (vCJD)*, sec. 5.25; World Health Organization, "Possible Creutzfeldt-Jakob Disease in

an Adolescent," *Weekly Epidemiological Record* 15 (1994): 105–106.

9   *beef burger was Vickie's favorite food:* A. Watkins, "Today Investigation," *Today* (London), January 13, 1994.

10   *"no evidence whatever that BSE causes CJD":* Phillips, Bridgeman, and Ferguson-Smith, "Emergence of Variant CJD," chap. 5 in *BSE Inquiry*, vol. 8, *Variant CJD (vCJD)*, sec. 5.29; see also footnote 220, citing "Statement of CMO on CJD," Department of Health press release, January 26, 1994 (YB94/1.26/3.1).

10   *coma that lasted four and a half years:* Phillips, Bridgeman, and Ferguson-Smith, "Statement No. 208: Beryl Rimmer," in *BSE Inquiry*.

10   *"ingestion of the infective agent may be one natural mode":* M. Kamin and B. M. Patten, "Creutzfeldt-Jakob Disease: Possible Transmission to Humans by Consumption of Wild Animal Brains," *American Journal of Medicine* 76 (1984): 142–145.

10   *minimized heredity or direct contact:* J. R. Berger, E. Weisman, and B. Weisman, "Creutzfeldt-Jakob Disease and Eating Squirrel Brains," *Lancet* 350, no. 9078 (1997): 642.

11   *Analysis of cells from the man's and the cat's brains showed remarkably similar abnormalities:* G. Zanusso et al., "Simultaneous Occurrence of Spongiform Encephalopathy in a Man and His Cat in Italy," *Lancet* 352, no. 9134 (1998): 1116–1117.

12   *"possibility that [these cases were] causally linked to BSE":* R. G. Will et al., "A New Variant of Creutzfeldt-Jakob Disease in the UK," *Lancet* 347, no. 9006 (1996): 921–925.

12   *indistinguishable from the agent that caused BSE:* A. F. Hill et al., "The Same Prion Strain Causes vCJD and BSE," *Nature* 389 (1997): 448–450.

13   *animals "lie down . . . and finally become lame":* Brown and Bradley, "1755 and All That," 1688.

13   *"malady of madness and convulsions":* H. B. Parry, *Scrapie Disease in Sheep: Historical, Clinical, Epidemiological, Pathological, and Practical Aspects of the Natural Disease* (New York: Academic Press, 1983).

13   *origins remained a mystery:* Brown and Bradley, "1755 and All That."

13   *scrapie agent remained infectious:* Centers for Disease Control and Prevention and National Institutes of Health, "Prions," in *Biosafety in Biomedical and Microbiological Laboratories*, edited by J. Y. Richmond and R. W. McKinney (Washington, DC: US Government Printing Office, 1999), sec. VII-D.

13   *a nonliving infectious agent:* Brown and Bradley, "1755 and All That."

14   *disease that was killing the Foré people:* Ibid.

14   *brains of kuru victims looked a lot like those of CJD victims:* Ibid.; R. B. Wickner et al., *Prions in Yeast Are Protein Genes: Inherited Amyloidosis* (Bethesda, MD: National Institutes of Health, National Institute of Diabetes and Digestive and Kidney Diseases, 2002); Karolinska Institute, "The 1997 Nobel Prize in Physiology or Medicine," press release, October 6, 1997.

14   *three brain-wasting diseases came to largely define TSEs:* Brown and Bradley, "1755 and All That."

14   *1976 Nobel Prize in Physiology or Medicine:* Karolinska Institute, "The 1976 Nobel Prize in Physiology or Medicine," press release, October 14, 1976.

15   *prions, do not reproduce:* S. B. Prusiner, "The Prion Disease."

15   *Scrapie-infected sheep remained the top suspect:* J. Hope et al., "Molecular Analysis of Ovine Prion Protein Identifies Similarities between BSE and an Experimental Isolate of Natural Scrapie, CH1641," *Journal of General Virology* 80 (1999): 1–4.

. . . .

15   *not a single documented case of a cow becoming sick from scrapie:* M. Balter, "On the Hunt for a Wolf in Sheep's Clothing," *Science* 287, no. 5460 (2000): 1906–1908.

15   *Nor . . . was a single case documented of a person becoming sick from the sheep disease:* Ibid.

15   *Perhaps a protein in a cow's brain had randomly mutated:* C. A. Donnelly, Department of Infectious Disease Epidemiology, Imperial College London, personal communication, November 26, 2002.

17   *renderers in France:* European Commission, *Report of the European Union with Regard to the Implementation of Council Directive 90/667/EEC and Commission Decision 91/516/ EEC concerning the Use of Prohibited Ingredients in the Animal Feedingstuff (19–20 August 1999),* ref. no. XXIV/1234/99 (Brussels: European Union, 1999). The report states: "Certain plants in the French rendering industry have used for years prohibited substances such as sludge from the biological treatment of the waste water or water from septic tanks from their own establishments or, possibly, from their suppliers."

18   *MBM. . . . is added to animal feed:* J. Lederberg, R. E. Shope, and S. C. Oaks Jr., eds., *Emerging Infections: Microbial Threats to Health in the United States* (Washington, DC: National Academies Press, 1992).

19   *solvents had little or no impact on the prion:* A. C. F. Colchester and N. T. H. Colchester, "The Origin of Bovine Spongiform Encephalopathy: The Human Prion Disease Hypothesis," *Lancet* 366, no. 9488 (2005): 856–861.

19   *dramatic increase in the number of sheep in the United Kingdom:* Department for Environment, Food and Rural Affairs (London), "Sheep and Lambs; Mutton and Lamb," table 5.14 in *Agriculture in the United Kingdom* (London: Department for Environment, Food and Rural Affairs, 2002).

19   *cows would have become infected before the 1980s:* Colchester and Colchester, "Origin of Bovine Spongiform Encephalopathy."

20   *BSE rapidly spread . . . via feed containing meat and bone meal of infected animals:* J. W. Wilesmith et al., "Bovine Spongiform Encephalopathy: Epidemiological Studies," *Veterinary Record* 123 (1988): 638–644.

20   *"a recipe for disaster":* Phillips, Bridgeman, and Ferguson-Smith, "Executive Summary of the Report of the Inquiry," chap. 1 in *BSE Inquiry*, vol. 1; see "Key Conclusions" in *Findings and Conclusions*.

20   *1988 ban on the feeding of recycled animal protein:* Donnelly, personal communication.

21   *an estimated 1 million cows had been infected:* Ibid.

21   *cases had been confirmed in more than 35,000 herds:* Centers for Disease Control and Prevention, "Questions and Answers Regarding Bovine Spongiform Encephalopathy (BSE) and Creutzfeldt-Jakob Disease (CJD)."

21   *in Belgium, Denmark, Switzerland, Italy, Greece, Germany, France, the Netherlands, Portugal, Ireland, and Spain:* Office International des Épizooties, "Number of Reported Cases of Bovine Spongiform Encephalopathy (BSE) Worldwide as of 04.03.2003" (Paris: Office International des Épizooties, 2003); Centers for Disease Control and Prevention, "Update 2002: Bovine Spongiform Encephalopathy and Variant Creutzfeldt-Jakob Disease."

21   *125 cases of the human form of mad cow disease:* Centers for Disease Control and Prevention, "New Variant CJD: Fact Sheet."

21   *history of exposure . . . where the disease was occurring in cattle:* Ibid.

21   *Letter to Christopher Melani:* Photocopy of letter from

. . . .

Michael W. Miller to Christopher Melani, December 20, 1997.

23   *"What's done is done"*: Quoted in "Wasting Away in the West," *CBS Evening News*, March 13, 2001.

23   *by 2001 the disease had spread to Nebraska*: US Department of Agriculture, Animal and Plant Health Inspection Service, Veterinary Services, "APHIS Factsheet," October 2001.

23   *three young venison-eaters*: E. D. Belay et al., "Creutzfeldt-Jakob Disease in Unusually Young Patients Who Consumed Venison," *Archives of Neurology* 58, no. 10 (2001): 1673–1678.

24   *five reported vCJD cases*: Centers for Disease Control and Prevention, "New Variant CJD: Fact Sheet."

24   *more than worrisome*: Ibid.

24   *victims' median age was sixty-eight*: Ibid.; R. V. Gibbons et al., "Creutzfeldt-Jakob Disease in the United States: 1979–1998," *Journal of the American Medical Association* 284, no. 18 (2000): 2322–2323.

25   *"more likely to get run over by a Winnebago"*: T. Thorne, personal communication, July 2002.

25   *"the whole issue of prion disease . . . has to be confronted seriously"*: S. Rampton, "What about Mad Deer Disease?" *E Magazine* 12, no. 4 (July–August 2001).

25   *brains of the young victims*: P. Yam, "Shoot This Deer," *Scientific American* 288, no. 6 (June 2003): 38–43.

26   *chronic wasting disease . . . is as infectious to human tissue as BSE*: G. J. Raymond et al., "Evidence of a Molecular Barrier Limiting Susceptibility of Humans, Cattle, and Sheep to Chronic Wasting Disease," *EMBO Journal* 19 (2000): 4425–4430.

26   *a cow was experimentally infected*: C. Q. Choi, "Chronic Wasting Studies Announced," *Scientist*, November 8, 2002.

26   *new studies to determine CWD's contagiousness*: Ibid.

26 *officials began shooting deer from helicopters:* S. Blakeslee, "Clues to Mad Cow Disease Emerge in Study of Mutant Proteins," *New York Times*, May 23, 2000; S. Blakeslee, "Weighing 'Mad Cow' Risks in American Deer and Elk," *New York Times*, February 23, 1999.

26 *more than 1,000 Wisconsin deer had tested positive for CWD:* Wisconsin Department of Natural Resources, "Wisconsin's Chronic Wasting Disease Response Plan: 2010–2025" (Madison: Wisconsin Department of Natural Resources, 2010).

27 *science lost two of its greatest CWD researchers:* Associated Press, "National News: CWD Researchers Killed in Snowy Highway Crash," *Billings Gazette*, December 29, 2004.

28 *deer can shed the prions in their feces long before they show signs of the disease:* J. O'Brien, "National News: Prions Found in Feces of Deer Asymptomatic for Chronic Wasting Disease," University of California, San Francisco, September 9, 2009.

28 *a major study on the possibility of people contracting CWD:* J. Y. Abrams et al., "Travel History, Hunting, and Venison Consumption Related to Prion Disease Exposure, 2006–2007 FoodNet Population Survey," *Journal of the American Dietetic Association* 111 (no. 6) (June 2011): 858–863, doi:10.1016/j.jada.2011.03.015.

29 *banned the practice of feeding animal by-products to cattle:* Donnelly, personal communication.

29 *"highly unlikely that a person would contract vCJD . . . in the United States":* Food and Drug Administration, Center for Food Safety and Applied Nutrition, "Consumer Questions and Answers about BSE," March 2001.

30 *"Human remains are known to be incorporated into meal made locally":* Colchester and Colchester, "Origin of Bovine Spongiform Encephalopathy."

30   *How the cow contracted BSE is unknown:* Centers for Disease Control and Prevention, "BSE (Bovine Spongiform Encephalopathy, or Mad Cow Disease)."

Chapter 2. A Chimp Called Amandine: HIV/AIDS

33   *"the earliest beginnings of the world":* J. Conrad, "Heart of Darkness," in *The Portable Conrad,* edited by M. D. Zabel (New York: Viking Press, 1950), 536.

35   *a new immunosuppressive disease . . . had emerged:* M. S. Gottlieb et al., "Pneumocystis Pneumonia—Los Angeles," *Morbidity and Mortality Weekly Report* 30, no. 21 (1981): 250–252.

36   *the name "acquired immunodeficiency syndrome"—AIDS— was coined:* Centers for Disease Control and Prevention, "Where Did HIV Come From?"

36   *immune-ravaged patients in Europe:* M. D. Grmek, *History of AIDS: Emergence and Origin of a Modern Pandemic* (Princeton, NJ: Princeton University Press, 1990).

36   *researchers in the United States discovered a frozen blood sample:* M. Balter, "Virus from 1959 Sample Marks Early Years of HIV," *Science* 279, no. 5352 (1998): 801; A. G. Motulsky, J. Vandepitte, and G. R. Fraser, "Population Genetic Studies in the Congo: I. Glucose-6-Phosphate Dehydrogenase Deficiency, Hemoglobin S, and Malaria," *American Journal of Human Genetics* 18, no. 6 (1966): 514–537.

36   *man was from Léopoldville, Belgian Congo:* T. Zhu et al., "An African HIV-1 Sequence from 1959 and Implications for the Origin of the Epidemic," *Nature* 391, no. 6667 (1998): 594–597.

36   *earliest documented case of HIV-1 infection:* Balter, "Virus from 1959 Sample."

39   *an SIV . . . almost indistinguishable from HIV-2:* V. M. Hirsch

et al., "An African Primate Lentivirus (SIVsm) Closely Related to HIV-2," *Nature* 339, no. 6223 (1989): 389–392; R. V. Gilden et al., "HTLV-III Antibody in a Breeding Chimpanzee Not Experimentally Exposed to the Virus," *Lancet* 327, no. 8482 (March 22, 1986): 678–679.

40   *died from complications of childbirth:* "Pathology Worksheet (Necropsy): Animal I.D. Number 205" (Alamogordo: New Mexico State University, Primate Research Institute, 1985).

41   *They named her Amandine:* Martine Peeters, personal communication, June 2002.

41   *a series of seizures:* J. C. Vié et al., "Megaloblastic Anemia in a Handreared Chimpanzee," *Laboratory Animal Science* 39, no. 6 (1989): 613–615.

42   *the test came back positive:* M. Peeters et al., "Isolation and Partial Characterization of an HIV-Related Virus Occurring Naturally in Chimpanzees in Gabon," *AIDS* 3, no. 10 (1989): 625–630.

42   *Macolamapoye:* M. Peeters, personal communication, June 2002.

43   *Delaporte, a physician, was plenty worried:* E. Delaporte, personal communication, June 2002.

44   *virus fragments found in chimps were indeed related to HIV-1:* T. Huet et al., "Genetic Organization of a Chimpanzee Lentivirus Related to HIV-1," *Nature* 345, no. 6273 (1990): 356–359.

45   *researchers developed a method for identifying evidence of the virus in fecal samples:* B. F. Keele et al., "Chimpanzee Reservoirs of Pandemic and Nonpandemic HIV-1," *Science* 313, no. 5786 (July 28, 2006): 532, doi:10.1126/science.1126531.

47   *virus probably originally arose between 1900 and 1920:* P. M. Sharp and B. H. Hahn, "AIDS: Prehistory of HIV-1," *Nature* 455 (October 2, 2008): 605–606, doi:10.1038/455605a.

47   *infected chimps lived in the southeastern corner of Cameroon:* Ibid.

48   *"everyone's nightmare":* L. K. Altman, "H.I.V. Linked to a Subspecies of Chimpanzee," *New York Times,* February 1, 1999.

48   *a type of HIV derived from gorillas:* J.-C. Plantier et al., "A New Human Immunodeficiency Virus Derived from Gorillas," *Nature Medicine* 15 (2009): 871–872, doi:10.1038/nm.2016.

48   *"simplest explanation for how SIV jumped to humans":* Sharp and Hahn, "Prehistory of HIV-1."

49   *"one of their main sources of food is bushmeat":* Peeters, personal communication; see also M. C. Peeters et al., "Risk to Human Health from a Plethora of Simian Immunodeficiency Viruses in Primate Bushmeat," *Emerging Infectious Diseases* 8, no. 5 (2002): 451–457.

Chapter 3. The Travels of Antibiotic Resistance: *Salmonella* DT104

55   *The first, in 1996, struck nineteen schoolchildren:* R. G. Villar et al., "Investigation of Multi-Resistant *Salmonella* Serotype *typhimurium* DT104 Infections Linked to Raw-Milk Cheese in Washington State," *Journal of the American Medical Association* 281, no. 19 (1999): 1811–1816.

55   *people who had eaten unpasteurized Mexican-style soft cheese:* S. H. Cody et al., "Two Outbreaks of Multi-Resistant *Salmonella* Serotype *typhimurium* DT104 Infections Linked to Raw-Milk Cheese in Northern California," *Journal of the American Medical Association* 281, no. 19 (1999): 1805–1810.

55   *resistant to twelve different antibiotics:* D. G. White et al., "The Isolation of Antibiotic-Resistant *Salmonella* from Retail Ground Meats," *New England Journal of Medicine* 345, no. 16 (October 18, 2001): 1147–1154, doi:10.1056/NEJMoa010315.

56   *pathologist Theobald Smith:* C. E. Dolman, "Theobald Smith, 1859–1934: A Fiftieth Anniversary Tribute," *American Society of Microbiology News* 50, no. 12 (1984): 577–580.

56   *genus of bacteria that lives in the intestines of many species:* R. K. Robinson, C. A. Batt, and P. Patel, eds., *Encyclopedia of Food Microbiology*, vol. 3, *Salmonella* (London: Academic Press, 2000).

60   *additives to help pigs, chickens, and livestock gain weight more quickly:* L. Groeger, "A History of FDA Inaction on Animal Antibiotics," *ProPublica*, April 4, 2012.

61   *by 1963 type 29 had become resistant to two antibiotics:* E. S. Anderson and M. J. Lewis, "Drug Resistance and Its Transfer in *Salmonella typhimurium*," *Nature* 206, no. 4984 (1965): 579–583.

61   *rare forms had armed themselves against seven antibiotics:* Ibid.; E. S. Anderson, "Drug Resistance in *Salmonella typhimurium* and Its Implications," *British Medical Journal* 3 (1968): 333–339.

61   *Of some five hundred confirmed human cases . . . six were fatal:* Anderson, "Drug Resistance in *Salmonella typhimurium*."

61   *the dealer . . . apparently committed suicide:* L. Ward, head of the salmonella research unit, Public Health Laboratory Service, personal communication, June 2002.

61   *outbreaks were "almost entirely of bovine origin":* Anderson, "Drug Resistance in *Salmonella typhimurium*."

61   *"the time has clearly come for a re-examination":* Anderson and Lewis, "Drug Resistance and Its Transfer in *Salmonella typhimurium*," p. 583.

61   *use of antibiotics to make animals grow faster "should be abolished altogether":* Editorial, "A Bitter Reckoning," *New Scientist* (January 4, 1968): 14–15.

62   *"Unless drastic measures are taken"*: J. Bower, "The Farm Drugs Scandal," *Ecologist* 1 (1970): 10–15.

62   *"there is ample data now in the literature"*: FDA report cited by T. H. Jukes in "Public Health Significance of Feeding Low Levels of Antibiotics to Animals," *Advances in Applied Microbiology* 16 (1973): 1–29; quote, p. 19.

62   *indiscriminate antibiotic use "favors the . . . development of single- and multiple-antibiotic-resistant bacteria"*: C. D. Van Houweling, US Food and Drug Administration, press briefing statement, January 31, 1972.

62   *licenses for use of antibiotics as growth promotants should be revoked*: US Food and Drug Administration, *Antibiotics in Animal Feeds: Information for Consumers* (Rockville, MD: US Food and Drug Administration, Center for Veterinary Medicine, 1993); M. Mellon, "Antibiotic Resistance: Causes and Cures" (speech given to the National Press Club, Washington, DC, June 4, 1999).

62   *"a cult of food quackery"*: Jukes, "Public Health Significance."

62   *cited as evidence of the cult*: Ibid.

63   *advocated that antibiotics be routinely used in some human food*: Ibid.

63   *Antibiotics, he believed, could compensate for malnourishment*: Ibid.

63   *incidence of Salmonella type 29 declined*: E. J. Threlfall et al., "The Emergence and Spread of Antibiotic Resistance in Food-Borne Bacteria in the United Kingdom," *APUA Newsletter* (published by the Alliance for the Prudent Use of Antibiotics) 17, no. 4 (1999): 1–7.

64   *spread to Cambridgeshire and Yorkshire*: Editorial, "Why Has Swann Failed?" *British Medical Journal* 1, no. 6225 (1980): 1195–1196.

64   *"man who threw himself out of the Empire State Building"*: R.

Young et al., *The Use and Misuse of Antibiotics in UK Agriculture. Part 2: Antibiotic Resistance and Human Health* (Bristol, England: Soil Association, 1999).

64 *pharmaceuticals that carried a higher profit margin:* Ibid.

65 *half a dozen weighty scientific evaluations:* R. A. Stallones, "Epidemiology and Public Policy: Pro- and Anti-Biotic," *American Journal of Epidemiology* 115, no. 4 (1982): 485–491.

65 *outspoken advocate for unlimited antibiotic use:* P. B. Lieberman and M. G. Wootan, *Protecting the Crown Jewels of Medicine: A Strategic Plan to Preserve the Effectiveness of Antibiotics* (Washington, DC: Center for Science in the Public Interest, 1998).

65 *"If the decision were mine":* Stallones, "Epidemiology and Public Policy," 490.

66 *solved the case without using food as the witness:* T. F. O'Brien et al., "Molecular Epidemiology of Antibiotic Resistance in *Salmonella* from Animals and Human Beings in the United States," *New England Journal of Medicine* 307 (1982): 1–6.

66 *issue "certainly has been studied sufficiently":* B. Keller, "Ties to Human Illness Revive Move to Ban Medicated Feed," *New York Times*, September 16, 1984.

66 *Britain's ban of certain antibiotics . . . suggested exactly the opposite:* Ibid.

66 *budget . . . was in the hands of the same appropriations subcommittee:* Ibid.

67 *"use of any antimicrobial agent for growth promotion in animals should be terminated":* World Health Organization, "The Medical Impact of Antimicrobial Use in Food Animals: Report of a WHO Meeting, Berlin, Germany, 13–17 October 1997," WHO/EMC/ZOO/97.4.

68 *Bactrim was another option:* F. J. Angulo et al., "Origins and Consequences of Antimicrobial-Resistant Nontyphoidal *Sal-*

monella: Implications for the Use of Fluoroquinolones in Food Animals," *Microbial Drug Resistance* 6, no. 1 (2000): 77–83; M. Vijups, personal communications, July 2002.

69   *it struck seven people in Airdrie, Scotland:* Threlfall et al., "Emergence and Spread of Antibiotic Resistance"; R. Davies, *Zoonose-Nyt* 8, no. 1 (2001).

71   *resistant genes in DT104 have been traced back to fish bacteria:* F. Angulo, "Antimicrobial Agents in Aquaculture: Potential Impact on Public Health," *APUA Newsletter* 18, no. 1 (2000): 1, 4–5; P. Smith, "Aquaculture and Florfenicol Resistance in *Salmonella enterica* Typhimurium DT104," letter, *Emerging Infectious Diseases* 14, no. 8 (August 2008): 1327–1328, doi:10.3201/eid1408.080329; F. C. Cabello, "Aquaculture and Florfenicol Resistance in *Salmonella enterica* Serovar Typhimurium DT104," letter, *Emerging Infectious Diseases* 15, no. 4 (April 2009): 623, doi:10.3201/eid1504.081171; World Health Organization, "Antimicrobial Use in Aquaculture and Antimicrobial Resistance: Report of a Joint FAO/OIE/WHO Expert Consultation on Antimicrobial Use in Aquaculture and Antimicrobial Resistance, Seoul, Republic of Korea, 13–16 June 2006."

72   *"bacteria that lived on farmed fish in Southeast Asia":* E. H. Kim and T. Aoki, "Drug Resistance and Broad Geographical Distribution of Identical R Plasmids of *Pasteurella piscicida* Isolated from Cultured Yellowtail in Japan," *Microbiology and Immunology* 37, no. 2 (1993): 103–109.

72   *traced to fish meal from Peru:* G. M. Clark et al., "Epidemiology of an International Outbreak of *Salmonella agona*," *Lancet* 302, no. 7827 (September 1973): 490–493, doi:10.1016/S0140-6736(73)92082-5.

73   *hailed as a breakthrough treatment for many infections:* Lieberman and Wootan, *Protecting the Crown Jewels of Medicine*;

S. J. Olsen et al., "A Nosocomial Outbreak of Fluoroquino-lone-Resistant *Salmonella* Infection," *New England Journal of Medicine* 344 (2001): 1572–1579.

74 *began to show resistance to fluoroquinolones*: H. P. Endz et al., "Quinolone Resistance in *Campylobacter* Isolated from Man and Poultry Following the Introduction of Fluoroquinolones in Veterinary Medicine," *Journal of Antimicrobial Chemotherapy* 27 (1991): 199–208.

74 *licensed fluoroquinolone for treating and preventing illness in turkeys and chickens*: Angulo et al., "Origins and Consequences."

74 *Two years later, 16 percent . . . showed some resistance*: E. J. Threlfall, "Increasing Spectrum of Resistance in Multiresistant *Salmonella typhimurium*," *Lancet* 347, no. 9007 (1996): 1053–1054.

74 *by 1996, fluoroquinolone-resistant . . . infections were sickening people*: Threlfall et al., "Emergence and Spread of Antibiotic Resistance."

74 *full support of the CDC*: F. J. Angulo, personal communication.

75 *in 1995 the FDA granted approval*: Angulo et al., "Origins and Consequences"; S. Rossiter et al., "Emerging Fluoroquinolone Resistance among Non-Typhoidal *Salmonella* in the United States: NARMS 1996–2000" (presentation given at International Conference on Emerging Infectious Diseases, Atlanta, GA, March 26, 2002).

75 *By 2000, 1.4 percent of salmonella infections showed some resistance*: Rossiter et al., "Emerging Fluoroquinolone Resistance."

75 *strong evidence . . . that the use of fluoroquinolones in poultry posed a risk to human health*: Lieberman and Wootan, *Protecting the Crown Jewels of Medicine.*

75   *prompting the Bayer Corporation to launch a five-year battle*: Keep Antibiotics Working, "In Depth: Fluoroquinolones: Unnecessary Risks."

75   *"The consensus is that there is no public health risk"*: L. Fabregas, "Bayer, FDA Spar over Safety of Poultry Drug," *Pittsburgh Tribune-Review*, January 20, 2002.

75   *"there is no scientific evidence"*: Animal Health Institute, "Statement by Alexander S. Mathews, 'The Use of Antibiotics in Food-Producing Animals,'" press release, May 28, 1998.

75   *"There is no clear documentation"*: R. Carnevale et al., "Fluoroquinolone Resistance in *Salmonella*: A Web Discussion," *Clinical Infectious Diseases* 31 (2000): 128–130.

76   *"no conclusive evidence"*: R. Robinson, in *Lancaster (PA) New Era*, March 6, 2002.

76   *"if we are what we eat, we're healthier if they're healthier"*: S. Lerner, "Risky Chickens," *Village Voice*, November 28–December 4, 2001.

76   *echoed the warnings from the 1960s*: "Bitter Reckoning."

77   *wastewater contaminates streams, rivers and aquifers, and lakes and their shores*: D. W. Kolpin et al., "Pharmaceuticals, Hormones, and Other Organic Wastewater Contaminants in U.S. Streams, 1999–2000: A National Reconnaissance," *Environmental Science and Technology* 36, no. 6 (2002): 1202–1211; M. Meyer et al., *Occurrence of Antibiotics in Surface and Ground Water Near Confined Animal Feeding Operations and Waste Water Treatment Plants Using Radioimmunoassay and Liquid Chromatography/Electrospray Mass Spectrometry* (Raleigh, NC: US Geological Survey, 2000).

77   *wastewater treatment plants in Europe*: J. Raloff, "Drugged Waters: Does It Matter That Pharmaceuticals Are Turning Up in Water Supplies?," *Science News Online*, March 21, 1998.

77   *two lakes in Switzerland*: R. Hirsch et al., "Occurrence of Antibiotics in the Aquatic Environment," *Science of the Total Environment* 225 (1999): 109–118.

77   *sediments under fish farms*: Letter from Union of Concerned Scientists to the US Environmental Protection Agency urging limits on vital antibiotics in factory farm effluent, August 3, 2000.

77   *up to 80 percent of the antibiotics used in aquaculture ends up in the environment*: World Health Organization, "Antimicrobial Use in Aquaculture."

79   *illness was traced back to ground beef that the patients had bought from grocery stores*: F. J. Angulo et al., "Common Ground for the Control of Multidrug-Resistant *Salmonella* in Ground Beef," *Clinical Infectious Diseases* 42, no. 10 (2006): 1455–1462, doi:10.1086/503572.

79   *outbreak led to the recall of nearly 500,000 pounds of ground beef*: AboutLawsuits.com, "Colorado Salmonella Outbreak Leads to Ground Beef Recall," July 23, 2009. Original recall notice: "Colorado Firm Recalls Ground Beef Products Due to Possible *Salmonella* Contamination," US Department of Agriculture, Food Safety and Inspection Service, News Release FSIS-RC-039-2009, July 22, 2009.

80   *use of the drugs in livestock was threatening their effectiveness*: Keep Antibiotics Working, "Keep Cephalosporin Antibiotics Working: Coalition Urges FDA to Reissue Ban on Extralabel Use of Drug," June 14, 2011.

80   *the FDA later withdrew the proposal*: Alliance for the Prudent Use of Antibiotics et al., joint letter to M. Hamburg, MD, commissioner, Food and Drug Administration, June 9, 2011.

81   *The Farm Bureau went so far as to claim*: A. Kar, "Misdirections and Feints on Antibiotic Use in Livestock Operations: Plays from the Industry Playbook," August 21, 2012. Original

Farm Bureau source: M. Maslyn, executive director, public policy, American Farm Bureau Federation, letter to US Food and Drug Administration, Center for Veterinary Medicine, July 12, 2012.

82   *These included resistance genes for most major classes of antibiotics:* Y.-G. Zhu et al., "Diverse and Abundant Antibiotic Resistance Genes in Chinese Swine Farms," *Proceedings of the National Academy of Sciences Early Edition* (2012): 1–6.

82   *from the farmlands of California to the pig farms of China:* White et al., "Isolation of Antibiotic-Resistant *Salmonella* from Retail Ground Meats."

Chapter 4. Of Old Growth and Arthritis: Lyme Disease

83   *Harrison's purchase was but a tiny grove:* D. S. Wilcove, *The Condor's Shadow* (New York: Freeman, 1999). Although Wilcove does not speak of Harrison's purchase, he describes the vastness of the early eastern forests of which that purchase was a part.

84   *farmland quilted the region:* E. B. Stryker, *Where the Trees Grow Tall* (Franklin Township, NJ: Franklin Township Historical Society, 1963), 173; J. P. Snell, *History of Hunterdon and Somerset Counties New Jersey: With Illustrations and Biographical Sketches of Its Prominent Men and Pioneers* (Philadelphia, PA: Everts & Peck, 1881), microfilm.

84   *more than half of the . . . northeastern forests had been cut:* L. C. Irland, *The Northeast's Changing Forests* (Petersham, MA: Distributed by Harvard University Press for Harvard Forest, 1999).

84   *several organizations . . . purchased the land:* E. Stiles, personal communication, June 2002.

87   *today forests cover three-quarters of their historical range:* Irland, *Northeast's Changing Forests.*

87 *Farms flowed in . . . and then washed out:* Wilcove, *Condor's Shadow*.

88 *white-tailed deer were almost extirpated:* L. Ragonese, "Deer Thrive as the Forests Sicken," *New Jersey Star-Ledger*, October 20, 1999. The article states that white-tailed deer were "nearly extinct in New Jersey 100 years ago."

92 *refuge and breeding grounds for mice and chipmunks:* K. A. Orloski et al., "Emergence of Lyme Disease in Hunterdon County, New Jersey, 1993: A Case-Control Study of Risk Factors and Evaluation of Reporting Patterns," *American Journal of Epidemiology* 147, no. 4 (1998): 391–397.

94 *average temperature in nearby New Brunswick increased:* US Environmental Protection Agency, *Climate Change and New Jersey* (Washington, DC: US Environmental Protection Agency, Office of Policy, Planning and Evaluation, 1997). The report states that "over the last century, the average temperature in New Brunswick, New Jersey, has increased from 50.4 degrees F (1889–1918 average) to 52.2 degrees F (1966–1995 average), and precipitation in some locations in the state has increased by 5–10%."

94 *an explosion in tick populations:* M. L. Wilson, "Distribution and Abundance of *Ixodes scapularis* (Acari: Ixodidae) in North America: Ecological Processes and Spatial Analysis," *Journal of Medical Entomology* 35, no. 4 (1998): 446–457; E. Lindgren and R. Gustafson, "Tick-Borne Encephalitis in Sweden and Climate Change," *Lancet* 358, no. 9275 (2001): 16–18. It should be noted that some scientists attribute the spread of ticks primarily to increasing deer populations.

100 *a cascade effect:* R. S. Ostfeld et al., "Effects of Acorn Production and Mouse Abundance on Abundance and *Borrelia burgdorferi* Infection Prevalence of Nymphal *Ixodes scapularis* Ticks," *Vector Borne and Zoonotic Diseases* 1, no. 1 (2001):

55–63; R. S. Ostfeld, "The Ecology of Lyme-Disease Risk," *American Scientist* 85 (1997): 338–346.

104 *third-highest number of Lyme disease cases:* R. Ostfeld, personal communication, November 2002.

104 *mice, not deer, were the strongest link:* R. S. Ostfeld et al., "Climate, Deer, Rodents, and Acorns as Determinants of Variation in Lyme-Disease Risk," *PLoS Biology* 4, no. 6 (2006): e145, doi:10.1371/journal.pbio.0040145.

106 *areas with more species had fewer cases of Lyme disease per capita:* R. S. Ostfeld and F. Keesing, "The Function of Biodiversity in the Ecology of Vector-Borne Zoonotic Diseases," *Canadian Journal of Zoology* 78 (2000): 2061–2078.

109 *indigenous forest dwellers might . . . have developed immunity to it:* J. A. Patz et al., "Effects of Environmental Change on Emerging Parasitic Diseases," *International Journal for Parasitology* 30, no. 12–13 (2000): 1395–1405.

Chapter 5. A Spring to Die For: Hantavirus

111 *"mysterious illness that killed two young Navajos":* Excerpted from Steve Sternberg, "An Outbreak of Pain," *USA Today*, July 2, 1998 (used by permission).

114 *The Colorado Plateau:* H. D. Grissino-Mayer, University of Tennessee, Knoxville, personal communication, May 2003; R. D. D'Arrigo and G. C. Jacoby, "A 1000-Year Record of Winter Precipitation from Northwestern New Mexico, USA: A Reconstruction from Tree-Rings and Its Relation to El Niño and the Southern Oscillation," *Holocene* 1 (1991): 95–101; H. D. Grissino-Mayer, "A 2,129-Year Reconstruction of Precipitation for Northwestern New Mexico, USA," in *Tree Rings, Environment, and Humanity: Proceedings of the International Conference, Tucson, Arizona, 17–21 May, 1994,* edited by J. S. Dean, D.

M. Meko, and T. W. Swetnam (Tucson, AZ: Radiocarbon, 1996), 191–204.

114 *stretches across 130,000 square miles of southeastern Utah:* R. Wheeler, "The Colorado Plateau Region," in *Wilderness at the Edge: A Citizen Proposal to Protect Utah's Canyons and Deserts* (Salt Lake City: Utah Wilderness Coalition, 1990).

116 *the extremes and duration of the heated Pacific appear to be something new:* K. Lewis and D. Hathaway, "Analysis of Paleo-Climate and Climate-Forcing Information for New Mexico and Implications for Modeling in the Middle Rio Grande Water Supply Study" (Boulder, CO: S. S. Papadopulos & Associates, 2001).

116 *A powerful El Niño cycle began in 1991:* R. Merideth, *A Primer on Climatic Variability and Change in the Southwest* (Tucson: University of Arizona, Udall Center for Studies in Public Policy and Institute for the Study of Planet Earth, 2001).

116 *quickly rising waters . . . sweeping a fifteen-year-old to his death:* Los Angeles County Office of Education and Department of Public Works, *No Way Out*, videotape produced by Nancy Rigg, 2000.

117 *winds laden with Pacific moisture pummeled Las Vegas:* K. Rogers, "Damage from Worst Flood Parallels Growth," *Las Vegas Review-Journal*, July 9, 1999.

117 *declared the normally arid state a flood disaster area:* Federal Emergency Management Agency, *Disaster Activity: January 1, 1992, to December 31, 1992* (Washington, DC: Federal Emergency Management Agency, 1992).

117 *series of mild snowstorms interspersed with rain:* Merideth, *Primer on Climatic Variability.*

117 *New Mexico was again declared a flood disaster area:* D. M. Engelthaler et al., "Climatic and Environmental Patterns Associated with Hantavirus Pulmonary Syndrome, Four Corners

Region, United States," *Emerging Infectious Diseases* 5, no. 1 (1999): 87–94.

117 *extensive roots quickly soaked up the water:* A. T. Carpenter and T. A. Murray, "Element Stewardship Abstract for *Bromus tectorum* L. (*Anisantha tectorum* (L.) Nevski)" (Arlington, VA: Nature Conservancy, n.d.).

117 *Snakeweed also burst forth:* Wyoming Agricultural Experiment Station, "Species Fact Sheet: Snakeweed Grasshopper, *Hesperotettix viridis* (Thomas)" (Laramie: Wyoming Agricultural Experiment Station, 1994).

118 *a year after a devastating influenza epidemic struck:* A. Crosby, *America's Forgotten Pandemic: The Influenza of 1918* (Cambridge, England: Cambridge University Press, 1989); G. Bailey and R. G. Bailey, *A History of the Navajos: The Reservation Years* (Santa Fe, NM: School of American Research Press, 1986).

118 *unusually heavy winter and spring rains:* Centers for Disease Control and Prevention, "Navajo Medical Traditions and HPS," 2000.

120 *evergreen plant was used by some Native Americans:* A. B. Reagan, "Plants Used by the White Mountain Apache Indians of Arizona," *Wisconsin Archaeologist* 8 (1929): 143–161; A. F. Whiting, *Ethnobotany of the Hopi*, Bulletin no. 15 (1939; reprint, Flagstaff: Museum of Northern Arizona, 1966).

120 *a cold medication . . . could loosen the patient's mucus:* S. A. Weber and P. D. Seaman, *Havasupai Habitat: A. F. Whiting's Ethnography of a Traditional Indian Culture* (Tucson: University of Arizona Press, 1985).

123 *directly followed these increased densities of mice:* Centers for Disease Control and Prevention, "El Niño Special Report: Could El Niño Cause an Outbreak of Hantavirus Disease in the Southwestern United States?," 2000.

123 *"increased winter and summer rain is associated with outbreaks of hantavirus":* J. N. Mills et al., "Long-Term Studies of Hantavirus Reservoir Populations in the Southwestern United States: A Synthesis," *Emerging Infectious Diseases* 5, no. 1 (1999): 135–142.

123 *researchers from Johns Hopkins:* G. E. Glass et al., "Using Remotely Sensed Data to Identify Areas at Risk for Hantavirus Pulmonary Syndrome," *Emerging Infectious Diseases* 6, no. 3 (2000): 238–247.

124 *"In beauty and harmony it shall so be finished":* G. Hausman, *Meditations with the Navajo: Prayers, Songs, and Stories of Healing and Harmony* (Rochester, VT: Bear & Company, 2001).

124 *the disease was old, perhaps even ancient:* J. S. Cameron, "The History of Viral Haemorrhagic Fever with Renal Disease (Hantavirus)," *Nephrology Dialysis Transplantation* 16, no. 6 (2001): 1289–1290, doi:10.1093/ndt/16.6.1289.

124 *in 1978 a man, also from Utah, had died:* Ibid.

124 *Hantavirus was later identified as the cause:* J. H. Simmons and L. K. Riley, "Hantaviruses: An Overview," *Comparative Medicine* 52, no. 2 (April 2002): 97–110.

125 *ten times as many of the mice in 1993:* Centers for Disease Control and Prevention, "Tracking a Mystery Disease: The Detailed Story of Hantavirus Pulmonary Syndrome (HPS)," August 29, 2012.

125 *By the summer of 2002, a total of 318 cases. . . . More than a third of the victims died:* Centers for Disease Control and Prevention, "Hantavirus Pulmonary Syndrome—United States: Updated Recommendations for Risk Reduction," *Morbidity and Mortality Weekly Report* 51, no. RR-9 (July 26, 2002): 1–12.

125 *case count peaked again in 2000 and 2006:* Centers for Dis-

ease Control and Prevention, "Annual U.S. HPS Cases and Case-Fatality, 1993–2012," 2012.

125 *Yosemite National Park:* World Health Organization, "Global Alert and Response (GAR): Hantavirus Pulmonary Syndrome—Yosemite National Park, United States of America," September 4, 2012.

126 *Nine of the victims had stayed at the Signature Tent Cabins:* Centers for Disease Control and Prevention, "Outbreak of Hantavirus Infection in Yosemite National Park," 2012.

126 *Forty percent of the victims died:* Centers for Disease Control and Prevention, "Annual U.S. HPS Cases and Case-Fatality, 1993–2012."

Chapter 6. A Virus from the Nile

127 *hottest month ever recorded in the city:* Environmental Defense et al., *Global Warming: Early Warning Signs: The Impact of Global Warming in North America* (Cambridge, MA: Union of Concerned Scientists, 1999).

128 *the life-threatening phase of his mysterious illness had passed:* J. Robin, "Quotes: On the Evening of August 11," *Newsday,* September 14, 1999.

129 *mosquito bit him one August evening:* J. Robin, "Victim's Final Days: Family Shares Memories with Loved One," *Newsday,* September 8, 1999.

129 *driest stretch in more than a hundred years:* Environmental Defense et al., *Global Warming: Early Warning Signs*; National Climatic Data Center, "Monthly Surface Data for Kennedy International Airport, April–June," (Asheville, NC: National Climatic Data Center, 1999).

129 *heat . . . killed more than a hundred people from the Midwest to the East Coast:* W. K. Stevens, "Across a Parched Land, Signs of Hotter Era," *New York Times,* August 1, 1999.

129 *northern house mosquito . . . often thrives during droughts:*
   P. R. Epstein and C. Defilippo, "West Nile Virus and
   Drought," *Global Change and Human Health* 2, no. 2
   (2001): 105–107; J. F. Day, "Predicting St. Louis Encephalitis Virus Epidemics: Lessons from Recent, and Not
   So Recent, Outbreaks," *Annual Review of Entomology* 46
   (2001): 111–138.

130 *At dusk they fanned across the borough:* Day, "Predicting St.
   Louis Encephalitis Virus Epidemics."

131 *his liver and kidneys began to fail:* D. S. Asnis et al., "The West
   Nile Virus Outbreak of 1999 in New York: The Flushing
   Hospital Experience," *Clinical Infectious Diseases* 30, no. 3
   (2000): 413–418.

131 *first fatality of the mysterious disease:* Robin, "Victim's Final
   Days."

131 *three more patients with neurological symptoms:* Asnis et al.,
   "West Nile Virus Outbreak of 1999."

131 *official . . . flew to Queens to interview surviving patients:* US
   General Accounting Office, *West Nile Virus Outbreak: Les-
   sons for Public Health Preparedness* (Washington, DC: US
   General Accounting Office, 2000).

132 *"do everything we can to wipe out the mosquito population":*
   R. Howell, "Mosquito Coast: Encephalitis Kills 1, Sickens
   Dozens along East River," *Newsday*, September 4, 1999.

132 *warning residents . . . to remain inside:* Ibid.

132 *developed a paralyzing phobia of flying insects:* D. Morrison,
   "'Very Paranoid' in Queens, Living with Fear of Virus," *Newsday*, October 2, 1999.

132 *hundreds of crows had begun dying:* US General Accounting
   Office, *West Nile Virus Outbreak.*

133 *veterinarian . . . treated more than fifty ill crows:* S. Shapiro,
   "As the Crow Dies: A Bird in Distress: Crows' Deaths Tied to

Drought," *Newsday*, September 14, 1999; J. Charos, personal communication, July 2001.

133 *dead crows all over the base*: C. Kilgannon, "At Fort Totten and Elsewhere, Crows Dying Mysteriously," *New York Times*, Sunday, August 22, 1999.

133 *a passerby happened upon four dead pigeons*: P. Dickens, "The Discovery of the West Nile," *Newsday*, September 26, 1999.

133 *a captive cormorant, three Chilean flamingos, a pheasant, and a bald eagle also died*: Ibid.

133 *worst die-off of crows in thirty years*: J. Steinhauer, "Outbreak of Virus in New York Much Broader Than Suspected," *New York Times*, September 28, 1999.

134 *all but ruled out Triple E as the culprit*: T. McNamara, personal communication, August 2001.

134 *all the birds stricken . . . were native to the Western Hemisphere*: Ibid.

134 *challenging the CDC's diagnosis*: U.S. General Accounting Office, *West Nile Virus Outbreak*: Ibid.

135 *Winds gusted to thirty miles per hour*: "History for Eilat, Israel: October 17, 1998," Weather Underground.

135 *flock of 1,200 birds set down at Eilat, in Israel's southern tip*: M. Malkinson et al., "Intercontinental Transmission of West Nile Virus by Migrating White Storks," *Emerging Infectious Diseases* 7, no. 3 (suppl.) (2001): 540.

135 *Israel's government tested a number of wild storks and the geese*: H. Bin, "West Nile Fever in Israel 1999–2000: From Goose to Man," in *International Conference on the West Nile Virus* (White Plains: New York Academy of Sciences, 2001).

137 *agreed to run further tests on the samples*: US General Accounting Office, *West Nile Virus Outbreak*.

138 *this spoke volumes*: Ibid.

138 *the CDC issued a press release*: Centers for Disease Control

and Prevention, Office of Communication, "West Nile–like Virus in the United States," September 30, 1999.

139 *matched a sample isolated from the brain of the dead goose in Israel*: R. S. Lanciotti et al., "Origin of the West Nile Virus Responsible for an Outbreak of Encephalitis in the Northeastern United States," *Science* 286, no. 5448 (1999): 2333–2337.

139 *a strain common throughout the Middle East*: M. Giladi et al., "West Nile Encephalitis in Israel, 1999: The New York Connection," *Emerging Infectious Diseases* 7, no. 4 (2001): 659–661.

139 *More than 20 million overseas passengers disembark there annually*: Port Authority of New York–New Jersey, Aviation Department, "Monthly Summary of Airport Activities: August 1999" (New York: John F. Kennedy International Airport, 1999).

140 *In spring and early summer, the songs of warblers . . . fill the wetlands*: M. T. Fowle and P. Kerlinger, *The New York City Audubon Society Guide to Finding Birds in the Metropolitan Area* (Ithaca, NY: Cornell University Press, 2001).

140 *rare sighting of a European widgeon*: Audubon Christmas bird count, Queens, New York, 1998–1999.

141 *they congregate by the thousands before continuing their patient journeys*: G. Waldbauer, *Millions of Monarchs, Bunches of Beetles* (Cambridge, MA: Harvard University Press, 2000).

141 *seven of the fifty-nine people hospitalized . . . had died*: D. Nash et al., "The Outbreak of West Nile Virus Infection in the New York City Area in 1999," *New England Journal of Medicine* 344, no. 24 (2001): 1807–1814.

142 *numerous parrots, parakeets, or lovebirds smuggled through New York each year*: R. Lewis, "With Evidence of Lingering Virus in New York Mosquitoes, Investigators Focus on Preventing Possible Outbreak," *Scientist* 14, no. 8 (2000): 1.

142 *passed the infection to a bird that migrated to Queens:* J. H. Rappole, personal communication, November 2001; J. H. Rappole, S. R. Derrickson, and Z. Habalek, "Migratory Birds and Spread of West Nile Virus in the Western Hemisphere," *Emerging Infectious Diseases* 6, no. 4 (2000): 319–328, November 2001.

142 *migrated south along the Atlantic Flyway:* Rappole, Derrickson, and Habalek, "Migratory Birds and Spread of West Nile Virus."

143 *Many of these may have survived an infection:* Centers for Disease Control and Prevention, "West Nile Virus: Vertebrate Ecology," 2002.

143 *sixty-four-year-old woman . . . came down with the disease:* E. De Valle, "A Break in West Nile War as Second Case Found in State," *Miami Herald,* August 10, 2001.

143 *"the mosquitoes are here and will always be here":* S. Wiersma, personal communication.

143 *She recovered and was released from the hospital:* J. Babson, "Closer to Home: West Nile Infects Woman in Keys," *Miami Herald,* August 25, 2001.

143 *virus had infected people in ten eastern states:* L. K. Altman, "Four Are Killed in Big Outbreak of West Nile Virus on Gulf Coast," *New York Times,* August 3, 2002.

144 *potentially alerting people in the disease path:* E. Young, "West Nile Virus Will Sweep across Whole US," 2002; National Aeronautics and Space Administration, "Satellites vs. Mosquitoes: Tracking West Nile Virus in the U.S.," press release no. 02-029, 2002.

144 *identified Louisiana as a potential trouble spot:* D. Rogers, personal communication, July 2002.

144 *Fifty-eight people fell ill:* Altman, "Four Are Killed in Big Outbreak."

146 *the first human infection in Maine*: J. Farwell, "Cumberland County Man Maine's First Case of West Nile in a Human," *Bangor Daily News*, October 31, 2012.

146 *West Nile virus has changed the way of life for many*: A. M. Kilpatrick, "Globalization, Land Use, and the Invasion of West Nile Virus," *Science* 334, no. 6054 (October 21, 2011): 323–327, doi:10.1126/science.1201010.

147 *more than sixty species of mosquitoes and more than three hundred species of birds*: D. J. Gubler, "The Continuing Spread of West Nile Virus in the Western Hemisphere," *Clinical Infectious Diseases* 45, no. 8 (2007): 1039–1046, doi:10.1086/521911.

147 *Warmer temperatures, higher humidity, and heavy rain*: J. E. Soverow et al., "Infectious Disease in a Warming World: How Weather Influenced West Nile Virus in the United States (2001–2005)." *Environmental Health Perspectives* 117, no. 7 (July 2009): 1049–1052, doi:10.1289/ehp.0800487.

147 *Hill City, Kansas, reached a high of 115 degrees*: Associated Press, "Hill City Hits 115 Again Wednesday," *Kansas City Star*, June 27, 2012.

147 *floods, wildfires, and storms, with record storm surges*: L. Morello and ClimateWire, "2012 Proves Warmest Year Ever in U.S.," *Scientific American*, January 9, 2013; National Oceanic and Atmospheric Administration, National Climatic Data Center, "2012 Global Temperatures 10th Highest on Record," 2012.

147 *more than 2,800 infections . . . with half ending up in meningitis or encephalitis*: Centers for Disease Control and Prevention, "Final 2012 West Nile Virus Update."

147 *Texas . . . had more than 1,700 cases and 76 deaths*: Centers for Disease Control and Prevention, "West Nile Virus," 2012.

147 *the mayor of Dallas declared a state of emergency*: M. Fer-

nandez and D. G. McNeil Jr., "West Nile Hits Hard around Dallas, with Fear of Its Spread," *New York Times*, August 16, 2012.

148 *hot weather can increase the chances of infection in several ways:* W. K. Reisen, Y. Fang, and V. M. Martinez, "Effects of Temperature on the Transmission of West Nile Virus by *Culex tarsalis* (Diptera: Culicidae)," *Journal of Medical Entomology* 43, no. 2 (2006): 309–317; J. P. DeGroote et al., "Landscape, Demographic, Entomological, and Climatic Associations with Human Disease Incidence of West Nile Virus in the State of Iowa, USA," *International Journal of Health Geographics* 7 (May 1, 2008): 19, doi:10.1186/1476-072X-7-19.

148 *the robin . . . has become suspect number one in the spread of West Nile virus:* Kilpatrick, "Globalization, Land Use, and the Invasion of West Nile Virus."

149 *urbanization and climate change are key elements in the emergence of new disease:* Ibid.

149 *the CDC . . . offering for the first time direct grants to states and cities to study the health effects of climate change:* Centers for Disease Control and Prevention, "CDC Features: Public Health Response to a Changing Climate," April 15, 2013.

Chapter 7. Birds, Pigs, and People: The Rise of Pandemic Flus

152 *they hadn't been infected by a common source:* G. Neumann, T. Noda, and Y. Kawaoka, "Emergence and Pandemic Potential of Swine-Origin H1N1 Influenza Virus," *Nature* 459, no. 7249 (June 18, 2009): 931–939, doi:10.1038/nature08157.

152 *Mexican authorities reported an outbreak of severe respiratory disease:* Ibid.

153 *nearly one-third of the population of La Gloria:* Centers for

Disease Control and Prevention, "Update: Novel Influenza A (H1N1) Virus Infection—Mexico, March–May, 2009," *Morbidity and Mortality Weekly Report (MMWR)* 58, no. 21 (June 5, 2009): 585–589.

153 *Many victims were young adults:* Neumann, Noda, and Kawaoka, "Emergence and Pandemic Potential of Swine-Origin H1N1 Influenza Virus."

154 *a mysterious respiratory outbreak hit Fort Dix, New Jersey:* Anonymous, "'Swine Flu' Originated at Fort Dix," *Lakeland (FL) Ledger,* 1976.

154 *the CDC sent a memorandum . . . urgently recommending mass immunization:* W. R. Dowdle, "The 1976 Experience," *Journal of Infectious Diseases* 176 (1997): S69–S72.

154 *a crash program to "inoculate every man, woman and child in the United States":* A. Pollack, "Lessons from a Plague That Wasn't," *New York Times,* October 23, 2005.

154 *confirmed toll was 230 cases, with 13 hospitalizations and 1 death:* J. Lessler et al., "Transmissibility of Swine Flu at Fort Dix, 1976," *Journal of the Royal Society Interface* 4, no. 15 (August 22, 2007): 755–762, doi:10.1098/rsif.2007.0228.

155 *World Health Organization (WHO) Collaborating Centers:* Centers for Disease Control and Prevention, "Selecting the Viruses in the Seasonal Influenza (Flu) Vaccine," March 9, 2011.

156 *a pandemic strain took six to nine months to spread around the world:* World Health Organization, "World Now at the Start of 2009 Influenza Pandemic," statement to the press by WHO Director-General Dr. Margaret Chan, June 11, 2009.

156 *more than 2 million people typically fly from Mexico:* K. Khan et al., "Spread of a Novel Influenza A (H1N1) Virus via Global Airline Transportation," *New England Journal of Medicine* 361, no. 2 (July 9, 2009): 212–214, doi:10.1056/NEJMc0904559.

156 *"confirmed importations associated with travel to Mexico"*: Ibid.

156 *the first US victim of the disease:* D. Althaus and J. Moreno, "Toddler Flu Victim Was Grandson of Press Baron," *Houston Chronicle*, May 1, 2009.

157 *Fifty-five-year-old Mitchell Wiener, an assistant principal:* L. Robbins, "More City Schools Closed by Flu," *New York Times*, May 20, 2009.

157 *cases suddenly exploded into 8,500, with 72 deaths, in thirty-nine countries:* Neumann, Noda, and Kawaoka, "Emergence and Pandemic Potential of Swine-Origin H1N1 Influenza Virus."

157 *"We are all in this together":* World Health Organization, "World Now at the Start of 2009 Influenza Pandemic."

158 *virus had gene segments from bird, human, and pig flu viruses:* Centers for Disease Control and Prevention, "CDC Briefing on Public Health Investigation of Human Cases of Swine Influenza, April 23, 2009, 3:30 p.m. EST," press release.

158 *This new three-headed virus:* R. J. Garten et al., "Antigenic and Genetic Characteristics of Swine-Origin 2009 A(H1N1) Influenza Viruses Circulating in Humans," *Science* 325, no. 5937 (July 10, 2009): 197–201, doi:10.1126/science.1176225; G. J. D. Smith et al., "Origins and Evolutionary Genomics of the 2009 Swine-Origin H1N1 Influenza A Epidemic," *Nature* 459, no. 7250 (June 25, 2009): 1122–1125, doi:10.1038/nature08182.

158 *a previously healthy seventeen-year-old boy from Wisconsin:* A. P. Newman et al., "Human Case of Swine Influenza A (H1N1) Triple Reassortant Virus Infection, Wisconsin," *Emerging Infectious Diseases* 14, no. 9 (September 2008): 1470–1472, doi:10.3201/eid1409.080305.

159 *All eleven victims ultimately survived:* V. Shinde et al., "Triple-Reassortant Swine Influenza A (H1) in Humans in the

United States, 2005–2009," *New England Journal of Medicine* 360, no. 25 (June 18, 2009): 2616–2625, doi:10.1056/NEJ Moa0903812.

159 *veterinarian J. S. Koen:* W. Ma, R. E. Kahn, and J. A. Richt, "The Pig as a Mixing Vessel for Influenza Viruses: Human and Veterinary Implications," *Journal of Molecular and Genetic Medicine* 3, no. 1 (2009): 158–166, doi:10.4172 /1747-0862.1000028.

160 *wild waterfowl and seabirds:* Ibid.

161 *Expanding markets in cities . . . have created major microbio-logical thoroughfares:* Ibid.

162 *a housewife in England, who kept a duck house next to a pond:* J. Kurtz, R. J. Manvell, and J. Banks, "Avian Influenza Virus Isolated from a Woman with Conjunctivitis," *Lancet* 348, no. 9031 (September 28, 1996): 901–902, doi:10.1016/S0140 -6736(05)64783-6.

162 *Genetic analysis of the 2009 H1N1 outbreak:* Smith et al., "Origins and Evolutionary Genomics of the 2009 Swine-Origin H1N1 Influenza A Epidemic"; Garten et al., "Antigenic and Genetic Characteristics of Swine-Origin 2009 A(H1N1) Influenza Viruses Circulating in Humans."

163 *five-year-old Edgar Hernandez of La Gloria, Mexico:* J. Cohen, "Texan Alleges Mexican Pig Farm May Be Liable for Pregnant Wife's Death from Swine Flu," *Science Insider*, May 14, 2009; O. R. Rodriguez, "La Gloria: Swine Flu's Ground Zero?," Associated Press, April 28, 2009; L. B. Martinez, "Family Might Seek $1B in Flu Death," *Brownsville (TX) Herald*, May 13, 2009.

163 *later survey of the hog farm failed to turn up the virus:* S. Fainaru, "Mexicans Blame Industrial Hog Farms," *Washington Post*, May 10, 2009.

163 *it had been circulating undetected for a decade:* Smith et al.,

"Origins and Evolutionary Genomics of the 2009 Swine-Origin H1N1 Influenza A Epidemic."

163 *convinced that a Mexican hog-farming operation gave rise to the flu virus that killed his wife:* B. Walsh, "H1N1 Virus: The First Legal Action Targets a Pig Farm," *Time: Health and Family,* May 15, 2009; Centers for Disease Control and Prevention, "Novel Influenza A (H1N1) Virus Infections in Three Pregnant Women—United States, April–May 2009," *Morbidity and Mortality Weekly Report (MMWR)* 58, no. 18 (May 15, 2009): 497–501; Cohen, "Texan Alleges Mexican Pig Farm May Be Liable."

163 *the first American citizen to die from H1N1:* Centers for Disease Control and Prevention, "Novel Influenza A (H1N1) Virus Infections in Three Pregnant Women."

164 *"reasonable to expect that this area around La Gloria is 'ground zero' for the H1N1-2009 swine influenza virus":* Cohen, "Texan Alleges Mexican Pig Farm May Be Liable."

164 *"The new H1N1 virus has largely run its course":* World Health Organization, "H1N1 in Post-Pandemic Period," director-general's opening statement at virtual press conference, August 10, 2010.

164 *"the mildest pandemic on record":* R. Knox, "Flu Pandemic Much Milder Than Expected," National Public Radio, December 8, 2009.

164 *disproportionate impact on children and young adults:* A. M. Presanis et al., "The Severity of Pandemic H1N1 Influenza in the United States, from April to July 2009: A Bayesian Analysis," *PLoS Medicine* 6, no. 12 (2009): e1000207, doi:10.1371/journal.pmed.1000207.

164 *"very misleading to describe that as mild":* Knox, "Flu Pandemic Much Milder Than Expected."

165 *"likely to be only a fraction of the true number of the deaths":* F.

S. Dawood et al., "Estimated Global Mortality Associated with the First 12 Months of 2009 Pandemic Influenza A H1N1 Virus Circulation: A Modelling Study," *Lancet Infectious Diseases* 12, no. 9 (September 2012): 687–695, doi:10.1016 /S1473-3099(12)70121-4.

165 *80 percent of them adults under the age of sixty-five:* Ibid.

166 *of the eighteen people infected in Hong Kong during the initial outbreak, six died:* Centers for Disease Control and Prevention, "Isolation of Avian Influenza A(H5N1) Viruses from Humans—Hong Kong, May–December 1997," *Morbidity and Mortality Weekly Report (MMWR)* 46, no. 50 (December 19, 1997): 1204–1207.

166 *there may be some limited person-to-person transmission:* Centers for Disease Control and Prevention, "Avian Influenza: Current H5N1 Situation," 2008.

166 *H5N1 was isolated from a flock of sick geese in Guangdong Province:* K. S. Yee, T. E. Carpenter, and C. J. Cardona, "Epidemiology of H5N1 Avian Influenza," *Comparative Immunology, Microbiology and Infectious Diseases* 32, no. 4 (July 2009): 325–340.

167 *not until 1955 was it determined to be an influenza A virus:* D. E. Swayne and D. L. Suarez, "Highly Pathogenic Avian Influenza," *Revue Scientifique et Technique de l'Office International des Épizooties* 19, no. 2 (2000): 463–482.

167 *A 2004 outbreak in Canada:* US Department of Agriculture, Animal and Plant Health Inspection Services, *Highly Pathogenic Avian Influenza Response Plan: The Red Book*, September 2012.

167 *Avian influenza viruses have also caused periodic die-offs of seals:* National Oceanic and Atmospheric Administration, NOAA Fisheries, Office of Protected Resources, "Declaration of 2011 Pinniped Unusual Mortality Event in the Northeast."

168 *new human H5N1 infections emerged in Vietnam, followed by sporadic cases in Europe, Africa, and the Middle East*: Yee, Carpenter, and Cardona, "Epidemiology of H5N1 Avian Influenza."

168 *human cases . . . seemed to follow outbreaks in poultry*: Ibid.

168 *Dead magpies later found on the farm had been infected by the virus*: Ibid.

169 *Over 100 million domesticated birds were culled*: Ibid.

169 *three civet cats died of H5N1 in Vietnam*: Ibid.

170 *"a purely avian virus adapted directly to human-to-human transmission"*: H. Chen et al., "Establishment of Multiple Sublineages of H5N1 Influenza Virus in Asia: Implications for Pandemic Control," *Proceedings of the National Academy of Sciences* 103, no. 8 (February 21, 2006): 2845–2850, doi:10.1073/pnas.0511120103.

170 *a mutant version of a key viral protein from bird flu*: D. Grady and D. G. McNeil Jr., "Debate Persists on Deadly Flu Made Airborne," *New York Times*, December 26, 2011; H.-L. Yen and J. S. M. Peiris, "Virology: Bird Flu in Mammals," *Nature* 486, no. 7403 (June 21, 2012): 332–333, doi:10.1038/nature11192.

170 *A mere four mutations later*: M. Enserink and J. Cohen, "One of Two Hotly Debated H5N1 Papers Finally Published," *Science Now*, May 2, 2012.

171 *papers detailing the results were finally published in their entirety*: Ibid.; S. Herfst et al., "Airborne Transmission of Influenza A/H5N1 Virus between Ferrets," *Science* 336, no. 6088 (June 22, 2012): 1534–1541, doi:10.1126/science.1213362; M. Imai et al., "Experimental Adaptation of an Influenza H5 HA Confers Respiratory Droplet Transmission to a Reassortant H5 HA/H1N1 Virus in Ferrets," *Nature* 486, no. 7403 (2012): 420–428.

171 *The third pandemic begins in early April 2013:* Centers for Disease Control and Prevention, "Human Infections with Novel Influenza A (H7N9) Viruses," CDC Health Advisory CDCHAN-00344 (April 5, 2013).

172 *"probably facilitate binding to human-type receptors and efficient replication in mammals":* T. Kageyama et al., "Genetic Analysis of Novel Avian A(H7N9) Influenza Viruses Isolated from Patients in China, February to April 2013," *Eurosurveillance* 18, no. 15 (April 11, 2013): 20453.

172 *genetic makeup of H7N9:* Ibid.

172 *"I think we need to be very, very concerned":* D. Butler, "H7N9 Bird Flu Poised to Spread," *Nature News,* April 15, 2013, doi:10.1038/nature.2013.12801.

172 *H7N9 can silently infect birds:* Ibid.

173 *mutation that allows the virus to grow well at a temperature similar to that of the human upper respiratory tract:* T. Devitt, "New Bird Flu Strain Seen Adapting to Mammals, Humans," *University of Wisconsin–Madison News,* April 12, 2013.

Epilogue: MERS-CoV and Beyond

175 *"outlook with regard to microbial threats to health is bleak":* M. S. Smolinski, M. A. Hamburg, and J. Lederberg, eds., *Microbial Threats to Health: Emergence, Detection, and Response* (Washington, DC: National Academies Press, 2003), 245, 247.

176 *"infectious diseases threaten national security":* J. Evans, "Pandemics and National Security," *Global Security Studies* 1, no. 1 (Spring 2010): 100–109.

176 *a newly discovered coronavirus:* R. J. de Groot et al., "Middle East Respiratory Syndrome Coronavirus (MERS-CoV): Announcement of the Coronavirus Study Group," *Journal of Virology* 87, no. 14 (July 2013): 7790–7792, doi:10.1128/jvi.01244-13.

. . . .

177 *a coronavirus of the same family:* World Health Organization, "Global Alert and Response (GAR): Novel Coronavirus Summary and Literature Update—as of 8 May 2013."

177 *MERS-CoV has so far been much deadlier:* Centers for Disease Control and Prevention, "Middle East Respiratory Syndrome (MERS): Frequently Asked Questions and Answers; MERS Cases and Deaths, April 2012–Present."

177 *the virus was spreading between people in close contact:* World Health Organization, "WHO Press Statement Related to the Novel Coronavirus Situation," May 12, 2013.

178 *antibiotic-resistant disease . . . kills some 20,000 people every year in the United States:* United States Congress, Office of Technology Assessment, *Impacts of Antibiotic-Resistant Bacteria,* OTA-H-629 (Washington, DC: US Government Printing Office, September 1995).

Acknowledgments

I wish to thank the many researchers who gave interviews, corresponded with me, or read drafts of chapters. Their comments, suggestions, and corrections were invaluable. Although any remaining mistakes belong to me, my appreciation belongs to them for their efforts in helping translate complex subjects into these seven stories.

In particular, Tony McMichael offered valuable comments on the introduction; David Bee, Marcus G. Doherr, Christl A. Donnelly, Peter Stent, and Tom Thorne assisted with the chapter on bovine spongiform encephalopathy; Robert Cooper, Bill Cummins, Eric Delaporte, Sian Evans, Beatrice Hahn, Martine Peeters, and Jean-Christophe Vié helped with the chapter on HIV/AIDS; Fred Angulo, Karen Florini, Cynthia Hawley, Mara Vijups, Patrick Wall, and Linda Ward provided invaluable assistance with the salmonella chapter;

Linda and John Beckley, Jeff Dunbar, Richard S. Ostfeld, Sarah E. Randolph, and the late Ted Stiles helped with the chapter on Lyme disease; James Cheek, H. D. Grissino-Mayer, Ben Muneta, Robert Parmenter, Steve Sternberg, and Ron Voorhees helped with the hantavirus chapter; and John Charos, Jonathan Day, Tracey McNamara, Mertyn Malkinson, Bob Perschel, John H. Rappole, Joshua Robin, and Steve Wiersma lent invaluable assistance with the chapter on West Nile virus.

My many conversations with other scholars who are helping to unravel the connections between ecology and health were invaluable. These include A. Alonso Aguirre, Peter Daszak, Jim Else, Paul R. Epstein, Gretchen Kaufman, Stuart B. Levy, Michael McCally, Ted Mashima, Steve Osofsky, Jonathan A. Patz, Mary C. Pearl, and Mark Pokras.

I am blessed to have had caring and gifted mentors throughout my life. Thomas Konsler patiently taught me the art of raising honeybees when I was seven—an experience that galvanized my interest in science; Mrs. Hutton and Joseph Lalley of Gibbons Hall encouraged me at a time in my life when I most needed it; and Ron Bromley, Chuck Carter, Doc Embler, Pop Hollandsworth, and John L. Tyrer influenced my life enormously when I was a student at the Asheville School. Derek Sarty taught me that the natural sciences could be as artistically inspiring as the expressive arts. Jeremy Dole gave me my first professional writing assignment. George Archibald has encouraged my work for nearly twenty years. The late Sheila Moffat, David Sherman,

Al Sollod, and Chip Stem provided encouragement and support through my years in veterinary school. I am indebted to Michael Lerner and Scott McVay, who have taught me so much. The late Franklin Lowe was an inspiring mentor.

Index

234   SEVEN MODERN PLAGUES

. . . .

Heyer Hills Farm, 58–60, 77–79
HIV. *See* HIV/AIDS
HIV-1 virus, 36–37, 41–43, 45
HIV-2 virus, 38–39
HIV/AIDS, xii, xiv, xxiv–xxv, 35–36, 90;
    chimpanzees' genetic analysis for, 44–47; GAB2
    positive but possible non-transmittal of, 42–45;
    human transmission events of, 46–50; study for
    determining origins of, 37–39; vaccine research
    for prevention of, 40–41. *See also* simian
    immunodeficiency viruses
hogs, 65–66, 82, 149; China's disposal of, 161;
    farms, 162–163; as influenza virus flophouse,
    159–160. *See also* livestock
Hong Kong, 165–166, 167–168
Hong Kong flu, xx, 157
House of Lords, 64
HPAI (highly pathogenic avian influenza), 167
HPS. *See* hantavirus pulmonary syndrome
human immunodeficiency virus. *See* HIV-1 virus;
    HIV-2 virus; HIV/AIDS
humans, 95, 102, 116, 162, 167; BSE contracting
    risks in, 7–8; H5N1 bird flu infections of,
    166, 170; H7N9 virus infecting, 173; HIV
    transmission events to, 46–50; HPS from
    environmental disharmony of, 119; kuru
    transmitted by consuming brains of, 6, 14; triple
    reassortant virus detected in, 158–159
human-type receptors, 172
Hunterdon County, NJ: climatic changes in, 94;
    deer density and Lyme disease in, 92–94;
    development of, 89–90, 94–95, 97; diseases
    striking, 90–91

immunities, xxiii
"Incurable Disease Wiping Out Dairy Cows" (1987),
    6
Indian Health Service, 119
Indian Reservation, 111–112
indigenous peoples, xxii–xxiii
infections, xviii–xix, xx–xxi, 175–176
influenza, xi, 159–161, 162, 167. *See also specific
    flus*
influenza A, 151, 159–160
interspecies borders, 168
intravenous antibiotics, 104

Jaffe, Harold W., 48
Jamaica Bay Wildlife Refuge, 140
Japan, 169
John Hopkins University, 123
Joiners of America, 85
joint stiffness, 103–104
Jukes, Thomas H., 62–63

Karoline (Woody Florena's cousin), 113–114
Koen, J. S., 159
Korean hemorrhagic fever, 124

kuru, human form TSE, 6, 14

La Gloria, Mexico, 163–164
La Niña, 115–116, 129
Legionella bacteria, 91
Legionnaire's disease, xx, 90–91, 111
Lenni-Lenape Indians, 83
Levy, Stuart B., 82
Lewis, David, 154
Lipsitch, Marc, 164
Liu Jianlun, xv
live animal markets, 162
livestock, 4, 59–61, 73–74; antibiotics' usage control
    for, 61–67, 80–82; England's management ques-
    tioned regarding, 16; FDA guidance on antibi-
    otics for, 81; feed rendered out of animal waste,
    17–18, 29–30; fish meal in feed for, 72–73
Lobb, Richard L., 76
loggers, 34–35, 49
L-type, BSE, 30
Lyme disease, xxiv, 98–100, 103–105, 109, 146;
    biological diversity reducing, 107–108; bull's-
    eye rash indicator of, 96, 99; deer density and,
    92–94; facial paralysis from, 103; introduction
    of, 91–92; Ostfeld collecting data on, 104;
    preventing, 95–96; from ticks, 96, 99; as vector-
    borne illness, 91

mad cow disease, xxiv, 7–12, 18–19, 29, 178; BSE
    technical name for, 1–4; disease spreading of,
    xxi, 2, 5–6, 16, 20–21; epidemic proportions
    of, 5–6, 20–21; scrapie's linkage to, 13–16, 19;
    TSE class in brain wasting disease, xii, xxi, xxiv,
    5–6. *See also* Creutzfeldt-Jacob disease; Pitsham
    Farm syndrome; transmissible spongiform
    encephalopathies (TSEs)
magpies, 168
malaria, xviii–xix, 36
Malaysian pig industry, xxiii–xxiv
Manlutac, Anna Liza, 151
man-made carnivores, 18–20, 31
Marilyn, chimpanzee, 40–42, 44–46
Mathews, Alexander S., 75
MBM. *See* meat and bone meal
McEwan, Doug, 24, 25
McFeeley, Patricia, 114
McNamara, Tracey, 133–139
McNeill, William H., xxii, xxvi
meat, 58–59
meat and bone meal (MBM), 18–19
media, xxi–xxii
medical technology, 178
Melani, Christopher, 21–22, 23, 25–26
meningitis, 146–147
MERS-CoV virus, xx, 176–177
Mexico, 140–142, 152–153, 156–157, 162–164
mice, xxv, 87, 92–93, 95, 97–109, 117; data studies
    of, 122–123, 125; heavy rainfalls increasing